To

Hinemoa Raharuhi

and her Mokopuna

Whose wise counsel

and eloquence

Energised my writing

And enhanced my understanding

Of

The ways of Water

Also by John Archer:

Dirt Cheap: The Mud Brick Book

Improvisations: Traditional
low-cost Building Techniques

Grass Roots, Earth Builder's Companion

Building for Kids and Adventurous Adults

The Home Building Experience

Building a Nation:
A history of the Australian house

Collins Australian Do-it-yourself Manual

Poles Apart:
Pole Frame Building in
Australia and New Zealand

The Owner Builder's Companion:
the best of the Owner Builder Magazine

On the Water Front:
Making your water safe to drink

A History of Australian Transport

The Water Efficient Garden: A practical
and innovative guide – from planning
through to established gardens

Bad Medicine:
is the health-care system letting you down?

The Great Australian Dream:
the History of the Australian House

The Water You Drink

Your Home:
the Inside Story of the Australian House

Sydney on Tap

Australia's Drinking Water:
The Coming Crisis

Twenty Thirst Century

the future of
water in Australia

John Archer

PURE
WATER
PRESS

Published by Pure Water Press
41 Cornelian Road Pearl Beach 2256
Tel: (02) 43 415 149

Printed by McPhersons Printing Group
Distributed by Tower Books
Design by Justin Archer Design

Archer, John, 1941– .

     Twenty thirst century : the future of water in Australia.

     Bibliography.

     ISBN 0 646 44535 9.

     1. Water-supply - Australia. 2. Water quality - Australia.

     3. Water conservation - Australia. I. Title.

333.910994

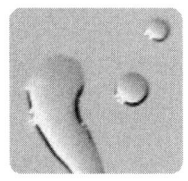

Contents

Acknowledgements

Twenty Thirst Century was handwritten (two drafts), edited, designed and produced during four intense months. This was made possible only by a sense of urgency and the encouragement and support of a dedicated group of friends and assistants.

Internet and documentary researcher Maya Verma was invaluable with her energy, enthusiasm, dedication and advice. Jo Thomson typed the manuscript, which was further enhanced by style editor Katrie Hilden. Laurel Fox-Allen, Simon Archer, Lucy Foenander and Cara Wilson, all provided valuable feedback on the final draft. Bulent Turkoglu, the Webmaster, designed and developed the website www.johnarcher.com.au to spread the news.

From beginning to end I was encouraged and nourished by Tramerag, the Healing Angel of the Mist, who was always there with her humour, sensitivity, perception, love, warm things to eat and endless faith in my ability.

Thank you all for everything you have done for me and for the waters.

Our gratitude is beyond measure.

John Archer

Preface

For the past fifteen years, I have been obsessed with water in all its forms and manifestations. Obsession was the only way I could come to grips with water's complexity, its sensitivity, its wisdom and, above all, its immortality.

My journey led me to remote temples in countries where water is worshipped as a living deity, to hot volcanic springs and icy waterfalls, to limpid pools hidden deep in the forest where I could swim with the yabbies and the frogs while cicadas and whipbirds sang in the sunset.

I have drunk the same water that dinosaurs drank in the Jurassic swamps, sucked the sweet dew from long grasstree flowers at dawn; I have sipped the sulphurous healing waters of Sukayu in the mountains of northern Japan and sung with other pilgrims in the legendary Golden Lotus tank at the Meenakshi Temple at Madurai in Tamil Nadu.

I have heard and recorded the stories, the legends and myths, intoned the hymns and prayers to the gods and goddesses of rain, learned the rituals, worshipped with reverence in holy places. For uncounted hours I have listened to the rhythms of the waves falling on the shore, to the sound patterns of lakes, small streams and majestic rivers.

Sometimes, watching the fluid reflections of the sun, moon and stars in the eternal moving mirror of the waters I have had the overwhelming sense of my consciousness merging with that of the infinite.

These then are my spiritual and environmental credentials.

Along with fifteen years of academic study, research and debate.

In this I have been ably supported by a small group of scientists, engineers, ecologists, lawyers and epidemiologists who have offered their friendship and invaluable expertise. Out of this research came five books on water issues in Australia. The success of these publications continues to fund my work.

My previous books were generated by a need to communicate a growing apprehension about the rapidly diminishing quantity and quality of our tap water, the water many of us drink, and its short- and long-term effects on our collective health and well-being.

Some readers found this information alarming and disturbing. That showed sound judgement: after all this is disturbing and alarming stuff. Politicians who felt threatened by my writing were quick to label me an alarmist. I make no apologies. When we see our own species moving rapidly and complacently towards ecological self-destruction, I believe it's our civic responsibility to sound the alarm.

My only concern is that not enough people seem to have heard.

However, it is encouraging to note that the ranks of celebrity alarmists in the scientific and academic communities seem to be increasing at a rapid (and unsustainable) rate in recent years.

And no wonder.

Massive and unprecedented environmental changes in the form of hailstorms, freak winds, floods, droughts, earthquakes, tsunamis and rising sea levels are already beginning to affect our lives. These are some of the manifestations of climate change we will confront during the Twenty Thirst Century.

Just as our past actions have precipitated the present situation, so too the options we choose now will affect our future.

We are at a point in our destiny where critical decisions are being made by governments and corporations, often without our knowledge or input, decisions which threaten to unravel once and for all the delicate web which up until now has supported our precarious existence.

In order to understand the magnitude and nature of that threat, you will need to equip yourself with enough information to find the tip of the iceberg, and where we stand in relation to it. Then you can form your own opinion, come to your own conclusions.

This book is offered in the hope that it may facilitate that process.

John Archer
Pearl Beach, 2005
www.johnarcher.com.au

Flying Blind
The future of water in Australia

As we race towards the future, we must never forget the fundamental reality of our situation: we are flying blind. Our dilemma is like that of a plane hurtling through the fog without a map or instruments. Instead of being able to provide a reliable radar system, scientists are peering through the cockpit window trying to warn of any obstacles ahead. And usually, the most they can say is that the dark mass looming into view might be a cloud bank.

Or then again, it might be a mountain.

> – Theo Colburn, Dianne Dumanoski
>   and John Peterson Myers, *Our Stolen Future*, 1996

Prepare for change.

We're not reducing our greenhouse gas emissions anything like fast enough to stop the effects of climate change this century. It's too late. We can slow warming down but we can't stop it. All the signs are that change will mean more extreme weather.

So if you think today's storms are wild, wait till global warming really kicks in.

You ain't seen nothing yet.

> – Gabrielle Walker, *New Scientist*, 16 September 2000

PICTURE THIS: IT'S SUNDAY – you're taking your family for a picnic. You're driving on the freeway. The speed limit is 110 km per hour, and you're right on it. A sign appears. It says 80. You keep on at 110. Further on another sign: 60. You take no notice. There's a man with a flag. You speed up to 120.

Then you see it. A tanker has overturned. The road is covered with oil and you're heading downhill towards it at 120km per hour.

And the children are playing in the back seat!

How are you going to handle this situation that you've got yourself into?

That scenario sums up my perspective on the future of water in Australia as things stand at the moment.

We've ignored nature's warning signs for years. Instead of doing the rational thing by limiting our insatiable appetite for economic growth and urban expansion, our leaders and planners still have their feet firmly on the accelerator as we approach what many believe will be a very slippery downhill slope.

Forget about the fact that this same scenario is happening all over the world – forget about saving the planet. Let's be selfish for a minute. Let's just think about ourselves and our immediate future. Forget about those utopian plans for our water supply in 2020 or 2050, and stay focused on the rapidly approaching reality of 2006, 2007, 2008, 2009 and 2010.

What if the dry weather continues for four or five more years?

In the millennial history of droughts five or ten years is minuscule. In some areas droughts have lasted hundreds, thousands, even millions of years, but they all began as six- or seven-year droughts, just like the one we're experiencing now. Except that they went on and on, year after year.

Of course, no one imagines or anticipates that possibility. We watch the sky and hope and pray for rain. Just as *they* did thousands of years ago, those long-vanished civilisations whose remains now lie hidden beneath the desert sands. Like us, they too were optimistic and went about their daily lives with an undiminished sense of purpose and achievement. They too dreamed of the world of possibilities their children might one day inherit.

But nature always has the last word.

The rain did not fall and eventually the great cities were abandoned, and all for the lack of water, mighty empires crumbled into dust.

Is it a coincidence that two leading writers on the international water crisis have prefaced their books with *Ozymandias*, Percy Bysshe Shelley's epitaph for a vanished kingdom?

> I met a traveller from an antique land
> Who said: "Two vast and trunkless legs of stone
> Stand in the desert. Near them on the sand,
> Half sunk, a shattered visage lies, whose frown,
> And wrinkled lip, and sneer of cold command,
> Tell that its sculptor well those passions read
> Which yet survive, stamped on these lifeless things,
> The hand that mocked them, and the heart that fed.
> And on the pedestal these words appear:
> 'My name is Ozymandias, King of Kings:
> Look upon my works, ye Mighty, and despair!'
> Nothing besides remains. Round the decay
> Of that colossal wreck, boundless and bare
> The lone and level sands stretch far away".

And what does Australia have in common with those long-lost civilisations buried in the Mesopotamian Desert, the mighty kingdoms of Sumer, Akkad and Babylon?

There are two fundamental similarities: Firstly, a complete dependence on the irrigated agriculture of inappropriate land resulting in levels of rising salinity that finally rendered the land and much of the water useless, and secondly, these great cities exceeded the capacity of their depleted and contaminated water supplies to provide for their needs.

Several scientists in this country have been trying to tell us for years that our only hope for the long-term future of our land and water is to revegetate something like 80% of our existing farmland.

Nobody wants to hear this.

Yet history tells us that most irrigation-based societies eventually fail. The pressure to increase productivity, when pursued to the ultimate, inevitably ends up damaging the very ecosystem that it depends on for its survival, and in the final analysis it is the availability of fresh unpolluted water which will determine whether we rise or fall, prosper or perish.

At present all of the nation's environmental indicators point in only one direction: downhill. We and our water are in decline, and the worst is yet to come if we do not alter our present course.

Of course there is always the possibility that the realisation of impending disaster may yet bring us to our senses.

Like Scrooge grimly confronting his future prospects in Charles Dickens' *A Christmas Carol*, we need to take a moment or two to consider our options.

> Before I draw nearer to that stone to which you point," said Scrooge,
> "answer me one question. Are these the shadows of the things that
> Will be, or are they the shadows of the things that May be only?"
> Still the Ghost pointed downwards to the grave by which it stood.
> "Men's courses will foreshadow certain ends, to which, if persevered
> in, they must lead," said Scrooge. "But if the courses be departed
> from, the ends will change. Say it is thus with what you show me!"

To change course now could indeed reshape our destiny.

It means rethinking our entire economic, social and environmental priorities on a massive and unprecedented scale.

It's not likely to happen.

Our politicians are not up to it. It's too hard, too frightening, and the end results are too far into the future. Any plan or project that exceeds a politician's anticipated life span in Parliament is doomed.

What about the electorate? What hope there?

Eighty-five per cent of Australia's population is essentially urban. Currently the voting majority seem more focused on interest rates and property values. They only worry about water when it looks as if it might run out. They don't seem to want to know much about the bigger picture. The rural electorate, on the other hand, is passionately and even violently opposed to any new proposition that might complicate their already difficult situation. So the status quo looks set to continue.

We are told that a token sum of $500 million has been set aside to "save" the Murray River. We have yet to learn that, unlike politicians, nature is not susceptible to bribes. Money is not the answer to our environmental problems.

What we lack is a sense of national purpose – a sense of direction.

Paralysed by indecision, like a deer caught in the headlights of an approaching vehicle, we are facing the rising threats of resource depletion, climate change, increased pollution, drought and urban water shortfalls with no realistic plans for a sustainable future.

If, indeed, a sustainable future still remains a possibility.

Is it too late for us to learn the lessons of history?

Have we strayed so far away from our connection with the natural world that we can't see for ourselves what's really happening around us?

Can't we see, on our TVs, ancient glaciers and polar ice caps melting for the first time in hundreds of thousands of years? Don't we realise that our prospects for the future are melting with them?

Or have we just become comfortably numb, cocooned in our great cities, insulated from reality, content with the illusion of a technological utopia somewhere up ahead?

It's time to wake up.

The party's just about over. We need to do some serious thinking about what to do next – because the decisions we make now will determine whether the future we dream of is about to evaporate.

Australians are used to overspending. As individuals we have racked up vast sums of credit card and bank debt to fund what we have come to refer to as our "lifestyle".

And just as we have exceeded our financial limits, so too have we exceeded nature's capacity to underwrite that lifestyle we aspire to. Our water account is overdrawn, and nature is calling in the debt, just as banks do when your credit rating drops to nil.

Make no mistake, this is no ordinary drought. This is the beginning of the new millennium. The century of climate change.

The Twenty Thirst Century.

We should have seen it coming and made plans to deal with it long ago, but our politicians, and the compliant scientists who fall into line behind them, have lied to us for years so that we would think they were doing their job. This has left us on the brink of economic and environmental disaster without a reliable compass to guide us.

It doesn't really matter anymore.

What matters is that we realise the precarious position we are in.

Our capital cities have long exceeded the sustainable limits of their water supplies, but some are in a better position to survive than others. Of all of them, Sydney remains the most exposed. This is partly because of lack of planning, mismanagement, incompetence and, until recently, lack of political interest, and partly because of its sheer size.

With less than two years of poor-quality water left, Sydneysiders have been told that there is no impediment to the city's growth, that it's still possible to continue to expand and multiply while urban sprawl and commercial development continue unabated.

The NSW Government is gambling the future of the city on three as yet unproven solutions to its water shortage: recycling of sewage, tapping into groundwater and desalination.

Other cities and population centres are offering their citizens the same recipe for salvation but unfortunately, on the available evidence none of these options will provide a sustainable solution in the short term, particularly for a metropolis as large as Sydney.

The most desirable option – recycling treated wastewater from sewage treatment plants for use on domestic gardens or for flushing toilets – would require a second pipeline. The cost of re-plumbing a large city with a second pipeline rules out this possibility. To duplicate Sydney's mains network would take years and cost billions of dollars.

The other less attractive alternative in relation to recycling sewage is to add the treated effluent to the reservoirs that supply our drinking water. It is vitally important that we examine carefully the health implications of drinking recycled effluent before such decisions are made. The NSW and Victorian Governments have both rejected this proposition outright. And quite correctly, for reasons discussed in detail in the chapters that follow.

Returning to the example of Sydney. The city has no infrastructure capable of handling any substantial water-recycling scheme for domestic use. The NSW Premier has rejected the idea of adding recycled effluent to drinking water supplies, and a second pipeline is out of the question according to Sydney Water's Managing Director, David Evans. Other capital cities are in a similar situation.

Recycling is a possibility for the distant future, but not for the present.

So to the second option, groundwater. Well yes there *is* some ground-water under Sydney but most of it is contaminated. The large Botany aquifer is polluted with toxic chemicals – and it is not the only one.

Conservative estimates suggest that underneath our cities and towns some 80,000 waste dumps and contaminated sites lie buried. Effluent from these continues to pollute underground aquifers. [1]

And even if this water were drinkable, the cost of pumps and pipelines to supply a significant proportion of Sydney's drinking water would be considerable. And then there are the power bills.

Such a project could also take several years to implement.

The use of groundwater is a doubtful option at best for any of our capital cities with the exception of Perth, which has substantial reserves.

That leaves desalination.

In spite of politicians' professed enthusiasm for desalination as a magic wand that will solve all their problems, it remains problematical. This is, after all, a technology that has been widely used by the US military, Middle Eastern countries and remote islands for the past 50 years.

Enough time, you would think, to solve those teething problems, iron out the wrinkles and refine the technique.

Apparently not.

If desalination is such a simple answer to the world's water shortage, why are a billion people going to bed thirsty every night?

And why has the northern hemisphere's largest and most modern desalination facility, in Israel, failed to deliver on its promise?

And why are other US cities such as San Diego cancelling their plans to build desalination plants, while others are closing existing plants down?

The short answer is that, while the desalination of seawater is viable on a small scale, the bigger the plant, the greater the possibility of costly system failure. In addition, the power requirements for a large plant can consume a significant proportion of a city's electricity supply, leading to a dramatic rise in greenhouse gases which will in turn exacerbate the very water shortage that led to the need for desalination in the first place.

So even if, as the friendly corporate desalination salesfolk suggest, desalination plants are about to mushroom all over the world, and even if they *could* deliver the goods, the world would soon become dependent on a highly energy-intensive life-support technology for its continued growth, and eventually its survival.

Polluting the oceans with vast quantities of extra unnecessary waste at the same time as we increase our contamination of the upper atmosphere is not a good idea either.

Our sustained pollution of the planet has already accomplished a couple of major climatic marvels. We need to take these factors into consideration.

We have increased the amount of carbon dioxide in the air by about 25% during the past century, and this will double during the current century: we have also more than doubled the level of methane and added to the mix a soup of other gases.

That's miracle number one: **we have substantially altered the atmosphere of the earth**, which leads us to to marvel number two.

Scientists in Britain are now warning that the extra carbon dioxide being absorbed by the world's oceans is changing their chemical composition, at the same time as global warming and human activity are altering the speed and temperature of the deep ocean currents that modify the climate of both the northern and southern hemispheres.

Changing the chemical composition and pH of the oceans will have far-reaching consequences for marine life and all the organisms, animals and people who depend on the sea for food and sustenance. Altering the deep ocean currents could cause the northern hemisphere to freeze over, while increasing the likelihood of droughts and hurricanes in the south.

Do we really want to give this self-destructive energy another boost by adding billions of tonnes of warm salt concentrate contaminated with toxic chemicals from our desalination plants?

Do we dare?

Haven't we done enough damage already?

Nature's capacity to correct our mistakes, to absorb the unrelenting volume of humanity's wastes, is not infinite. There are limits and we have exceeded them. Now we must face the consequences of our past excesses.

Preserving the delicate balance of nature is a sacred trust.

Abuse that trust and all bets are off.

After the scales are tipped, we can expect the unexpected. Expect weather events that haven't occurred before.

Expect record droughts, floods, hailstorms, hurricanes, cyclones, tsunamis, earthquakes, you name it.

The dire predictions of Nostradamus or the *Book of the Apocalypse* pale into insignificance beside the awesome power of natural catastrophes such as the great wave that struck the east coast of Australia long before our cities were built.

The return of such a wave – which was ten times higher than the tsunami that caused such destruction in Asia – would expose our over-valued seaside real estate for the illusion it is.

However, the great wave was a once-in-500-year event, probably caused by a meteor landing in the Pacific. The real threat from the ocean today is the slow inexorable intrusion of the sea into our coastal cities, aquifers, estuaries and productive agricultural land.

Remember those melting glaciers and polar icecaps that appear now and then on the evening news. The world's remaining ice cover contains so much water that, if it were all to melt, the sea level would rise by 75 metres. Yes, 75 metres. This potential flood is stored in the Greenland ice sheet (if it melts it will raise the world's oceans by 7 metres), the West Antarctic ice sheet (another 7 metres) and East Antarctica (60 metres), with a smaller amount, perhaps half a metre, stored in the planet's alpine glaciers.

All of these giant iceblocks are melting as we watch nature's response to our greenhouse gases. As global temperatures climb, the speed of the melting will increase and the sea levels will continue to rise.

We like to think that the high water mark is a relatively fixed point, but it is not as constant as we imagine.

During the last interglacial period, around 1,230,000 years ago, high tides peaked at 7 metres above their current level, whereas at the height of the last Ice Age when much of the world's water was frozen, the sea level was a massive 100 metres below what it is today.

In the past, such peaks and troughs occurred during long periods of geological time. Now global warming has so accelerated the process of change that what once took thousands of years is now happening during a human lifetime. We do not yet know what this will mean to us.

All I can say is: expect the unexpected.

Too much salt water on the one hand, too little fresh water on the other. As S.T. Coleridge's ancient mariner observed in the poem of the same name: "Water water everywhere, nor any drop to drink".

Not a very cheerful forecast, is it?

I wish I could offer a more optimistic one. After more than fifteen years of trying to sustain my optimism about the possibility of change in the water business, I'm afraid I've almost lost it.

Fifteen years is a long time in the water industry. I've watched vision-aries and pragmatists come and go, scientists with hollow promises of a future technological fix, smiling corporate clones with their grand "mission statements" about how they were destined to grow the water business into the twenty-first century, and the politicians and planners who continued to fund the whole show while assuring us that everything was under control.

And inexorably, year after year – in spite of the warnings, the promises, the mission statements, the influx of private enterprise, the billions of dollars wasted on mountains of reports and endless scientific study and monitoring – in spite of all this, the condition of Australia's water supplies continues its steady and predictable decline.

So where do we go from here?

This is a big question – there's no quick fix, no easy answer.

But I believe that collectively we can change our direction if we decide that's what we really want to do.

That's the first decision we need to make.

God sends
a messenger

In our general distraction with the complications of managing even the here and now, we forget to focus on determining at what point the shock will come, and what we will do then.

We know for example that you can't go on killing off the species in an ecosystem indefinitely; at some point it will fail like a building in which the key beam gives way and the whole system collapses. Yet we keep pushing the limits... we keep forgetting about that inevitable point of shock where the water supply is abruptly gone – when the lake goes anoxic, the forest turns to desert or the river turns to dust...

   – Ed Ayres, *Worldwatch*, July/August 1998

If waves crash up against the beach, eroding dunes and destroying homes, it is not the awesome power of Mother Nature. It is the awesome power of Mother Nature as it has been altered by the awesome power of man, who has overpowered in a century the processes that have been slowly evolving and changing of their own accord since the earth was born.

   – Bill McKibbon, *The End of Nature*, 1990

WHILE WRITING THE INTRODUCTION to *Twenty Thirst Century*, I was interrupted by a knock on the door. I went downstairs. Standing there quietly was a well-dressed man in his early 40s accompanied by a small boy. Both were wearing coats and ties, even though the day was hot.

"We come with a message from God," said the man as he extracted some magazines from his polished black briefcase.

"Yes of course," I replied.

Now I'm not a Christian believer but one thing I'll say for the Jehovah's Witnesses – no matter where I've lived they've always managed to track me down. Whether it was Holloways Beach, north of Cairns, in 1962, Tamborine Mountain in south eastern Queensland (1969), Mallacoota on the Gippsland coast of Victoria (1979), Alice Springs (1983) or NSW's Brunswick Heads (1989), there they were knocking patiently at the door, always two of them, always polite and endlessly enthusiastic (some might say persistent) in their desire for me to hear the latest word from the Lord.

Their message hasn't changed much in forty years, and most people are familiar with it. Basically it boils down to a warning that the end of the world is coming, so we'd better mend our ways and repent of our evil deeds before it's too late.

I'm sorry to confess that I haven't always been polite to the Jehovah's Witnesses. Sometimes their timing was bad. Other times I just wasn't in the mood to hear about the world ending and sent them away, with *Awake* and *The Watchtower* still in their hands.

And here it was, January 2005, and here they were once more.

"Did you know that the world is coming to an end?" asked the man earnestly.

"Yes," I said, "I believe you're right."

"Then you'll want to read this," he said, offering the latest editions of *Awake* and *The Watchtower*. I took the magazines from him and listened to his story. When he finished I thanked them both and closed the door.

Then I looked down. On the cover of *Awake* (8 January, 2005) was a familiar satellite image. The title posed the question "Can the planet Earth be saved?"

This happened to be the same question I had been wrestling with for months.

On my desk was a pile of the latest glossy, full-colour consultation drafts, strategies and water initiatives from all around Australia. There was:

A Thirst for Change – Water Proofing Adelaide

Water for Today and Tomorrow
– Help Brisbane Become a Water Smart City

Meeting the Challenges – Securing Sydney's Water Future

The Victorian Government White Paper,
Securing our Water Future Together – Our Water, Our Future

Securing our Water Future
– A State Water Strategy for Western Australia.

Could one detect a common theme, I wondered? Could there be some concern about the possibility of an uncertain future?

In spite of their carefully crafted beauro-speak, these documents were all asking much the same question, but unlike *Awake* they maintained a cheerful optimism that offered the hope that with the right politicians at the wheel and the appropriate procedures in place, at the end of the day our future would most certainly be secure.

The copywriters responsible for these dangerously misleading documents could have learned a lot from the anonymous author of *Awake*'s "Can the Planet be Saved?"

Who didn't pull any punches.

The article points out that despite of 30 years of international conferences devoted to environmental issues, beginning with the UN Conference on Human Environment in 1972, the destruction of the natural world has continued unabated.

> Unfortunately this rich body of treaties, action plans and other instruments has not reversed global environmental decline," says David Hunter, Executive Director of The Center for International Environmental Law. In fact, adds Hunter, "virtually every major environmental indicator is worse today than it was in the 1992 UN Conference.

Unfortunately this also is true for Australia.

The article continues:

> Why such meagre progress after more than 30 years of addressing environmental issues? One reason is the need for economic growth. The nations' economies are driven by consumer spending. That requires businesses to produce and that in turn takes raw materials. It is a vicious circle in which the environment ends up the loser.

Awake talks about the disappearing forests – about how back in 1999 the experts estimated that two thirds of the forest of the Philippines would be gone in ten years; that in Brazil, an area of forest the size of a soccer field disappears every eight seconds (count to eight and think about it!); how the overwhelming majority of Africa's citizens, who use wood as a cooking fuel, are destroying their own environment simply to survive.

This was all familiar stuff but reading it again somehow gave it a sharper edge, a heightened imperative. After all, this is the sort of information that deserves a wider audience. So I was heartened by the fact that *Awake* prints 22,842,000 copies in 87 languages and hand-delivers them personally around the world. It's a print run few environmental writers could ever hope for.

Back to the issues under discussion.

Awake devotes a page to water, beginning with quotes from *Time* (2002) which state that more than a billion people do not have easy access to clean drinking water. Sometimes pollution is the reason. According to the *Le Figaro*, the rivers of France are in a very poor state of health due to the accumulation of nitrate fertilisers. "French rivers discharged 375,000 tons of nitrates in the Atlantic in 1999, almost twice as much as in 1985," the paper reported.

In Japan, Jutaka Une, head of a non-profit farm safety organisation, warns of the rapid rise in groundwater contamination due to the high levels of chemical fertilisers and pesticides used to maintain the rate of food production.

Fifty percent of Brazil's sewage flows into its rivers, creating a chronic shortage of safe clean drinking water. Cities like Sao Paulo now have to import water from a hundred kilometres away because their own rivers are poisoned.

Our rivers are not poisoned, we say. We don't like such emotive language. Our rivers, we say, are unfit for drinking purposes. But no matter how you say it, it amounts to the same thing.

And that's a point that is often forgotten when water issues are discussed in the boardrooms of water corporations. The world's diminishing water is only half of the problem. The main concern is that what's left is often grossly polluted.

Australia rates a mention in *Awake*'s fact-finding round-up of imminent global catastrophes:

> Much of Australia's water shortage stems from a process called salinization. For decades landowners were encouraged to clear their land to plant crops. With fewer trees and shrubs to soak up the ground water, water tables began to rise bringing with them thousands of tons of subterranean salt. "Some 2.5 million hectares of land are already affected by salinity," says Australia's Commonwealth Scientific and Industrial Research Organisation (CSIRO). Much of this is Australia's most productive agricultural land.
>
> Some believe that if Australian legislators had not chosen profit over public interest, the salinity problem might have been avoided. "Governments were told from as early as 1917 that wheat belt soils

were especially prone to salinity," says Hugo Bekle of Edith Cowan University in Perth, Australia. "The impact of clearing on stream salinity was publicised by the 1920s and its effect on a rising water table was accepted by the Agricultural Department by the 1930s. A major report was undertaken for the (Australian) Government in 1950 ... yet governments persistently ignored these warnings, dismissing scientists as prejudiced."

The same could be said about the degradation of Australia's rivers and the lack of effective planning for water supplies to provide for our future.

One could fill large bookcases with all the studies, action plans, and research reports decades old and gathering dust, that warn of the gradual decline and eventual collapse of our aquatic environment.

Their unheeded messages represent a massive waste of resources in terms of time, money and energy.

Scientist's warnings are seldom given much credence by planners and their political masters. Yet, for instance, when the 2004 Asian tsunami struck, much emphasis was put on the lack of a science-based early warning system that might have given some advance notice of approaching calamity.

In fact there was a natural early warning system in operation, but few could read the signs. After the tsunami it was discovered that many animals had moved to higher ground long before the wave hit. In Sri Lanka and Thailand, elephants rounded up their young and moved off into the mountains several hours in advance of the fatal moment.

Now, instead of expensive scientific monitoring equipment that is always prone to human and mechanical error, the Sri Lankans propose to rely on elephants and their behaviour as early warning indicators.

Natural warning systems are cheaper and more reliable than science-based methods. Take the example of the canary in the coalmine. One of the great dangers of coalmining is the leaking of highly poisonous and inflammable gases. In order to detect these, a canary in a cage was carried down the pit. If the canary stopped singing and keeled over, the miners knew it was time to get out fast!

Wouldn't it be wonderful if we had a similar early warning system to tell us when our environment and our water supplies were really under serious threat, something that could ring alarm bells when we were in grave and mortal danger.

The good news is that there *is* such an early warning system.

The bad news is that we have been ignoring it for years. It's a disturbing thought.

Just pretend for a moment that we are coal-blackened miners working in a pit deep beneath the earth, and suddenly our little pet canary keels over.

What do we do?

Do we ignore the poor dead bird, and continue digging away because if we stop work our income will be curtailed? Is that wise? Is that prudent? Above all, is it safe?

The short answer is we won't know until something happens – but we can't say we weren't warned.

How many canaries need to lie dead at our feet before we get the message?

During the past 24 hours, 130 plant or animal species have disappeared forever.

Yes, folks, it's happening that fast.

In the ten seconds it has taken you to read this line, another hectare of rainforest has vanished – about 8000 hectares every day.

And talking about canaries, how do you think a canary might react to the world's daily dose of 10 million tonnes of toxic chemicals?

I think if I were down there in the mine I'd start to worry. Time to pile into the lifts and head for safety, leaving the danger behind.

If only we could do the same.

If only we could all crowd into a great big spaceship and take off for some pristine planet somewhere out there in the universe.

But you and I know that's not going to happen. Like it or not we're here for the long haul, so we'd better get a more realistic perspective about our future prospects.

While Australians worry about whether they'll have enough water to keep their lawns green, other people in less fortunate circumstances have difficulty getting enough clean water to drink. One billion of them, in fact. About one-fifth of the world's population – and that number is increasing by the minute.

Yeah, yeah, I know you've heard all this stuff before, but what can you do about it? Everyone has to make a living and pay the bills somehow and if that means that there are winners and losers in the world, well then that's just too bad. But you see it's not that simple.

There are no winners in a war against nature.

American author Gore Vidal put it this way: "Think of the earth as a living organism that is being attacked by billions of bacteria whose numbers double every forty years. Either the host dies or the parasite dies, or both die."

Whether you agree with Vidal's prognosis or not, one thing is certain: the world's population *is* indeed doubling every forty years.

What implications does this have for our water supplies here in Australia? Much as we like to think of ourselves as the Lucky Country, an island continent removed from the horrors of war and famine and pestilence that plague our neighbours to the north and west, we remain connected, if not by our common humanity, then by our total dependence on water for our economic, environmental and personal survival.

But forget about the economic and environmental imperatives for a minute and just focus on the issue of personal survival.

Your personal survival and that of your children and your family.

The gladiator Demetrius, standing in the arena of the Colosseum shouted to the Roman Emperor Nero, "You may threaten me, but nature threatens you."

Nature threatens us with extinction.

Make no mistake – we are in grave danger, not from the imagined and distant threat of terrorism, but from a more immediate, implacable, and relentless enemy – our own insularity, greed and stupidity.

One thing we have at least begun to acknowledge: having recognised the limits of our capacity to understand the complex web that sustains us, we know that we cannot accurately predict nature's response to our persistent and ever-increasing demands. One would think that this would make us more cautious in our approach to the environment and less *gung ho* about our capacity and ability to "manage" rivers, streams and aquifers.

But no.

Politicians, planners and even sometimes scientists act with a confidence born of ignorance and arrogance. Instead of acknowledging the mistakes of the past and working to correct them, each fresh crop of decisive decision makers dismisses the previous errors of others and promises a whole new beginning, and the process of destruction proceeds as before.

As the editor of *Water* magazine wrote in June 1997:

> Creeping disasters – those brought about by incremental changes over a long period of time – can be insidious. They may be difficult to detect in their early stages and impossible to avoid if action is delayed. As they develop over time, even when the threat is identified, political action may not be timely".

The authors of the best-selling *Limits to Growth* (1972) issued a similar warning to the world thirty years ago. Back in the 1970s, this dedicated group of philosophers, scientists, economists and researchers put together a series of predictive scenarios based on their model of the evolving world in the twentieth century.

If the message of *Limits to Growth* could be condensed into one brief paragraph it would probably be this simple statement:

> If the present growth trends in world population, industrialization, pollution, food production and resource depletion continue unchanged, the limits to growth on this planet will be reached sometime within the next 100 years.

They did not realise quite how soon this would be.

In 1997, a scientific study by Mathis Wackernagel and others came up with the term, "ecological footprint", to describe the amount of land required to provide the natural resources consumed by the population of various nations, as well as those resources necessary to absorb their wastes.

When their research was collated and analysed, it showed that the earth's people had been living beyond sustainable limits since the late 1980s. In every year after that, global consumption has exceeded the capacity of the earth to regenerate or provide the necessary amount of natural resources, to sustain this level of consumption.

By 1999, human demand had exceeded nature's supply by 20% and was still rising fast.

The unsustainable growth of the world's ecological footprint suggests that the end of the road is somewhere up ahead, and the speed at which we are travelling means that we may miss the warning signs and go straight over the cliff into the abyss.

If we fail to take corrective action in time, we may find that the economy we worship with such reverence is in free fall, and our cherished vision of the material world will fall along with it.

Sydney's ecological footprint, the amount of land required to produce goods and services and contain waste is still growing rapidly.

Between 1998 and 2003 for instance it grew by 23%, although population growth over the same period was only 7%. It now takes 7.4 hectares of land to maintain each Sydney resident's lifestyle. Obviously this growth cannot and will not continue indefinitely. Sooner or later, exceeding environmental limits will precipitate a catastrophic situation that will facilitate a sudden return to environmental equilibrium.

This could require **negative** growth of up to 20%. It will not be an easy transition.

In 2005, *Limits to Growth: The 30 Year Update* re-examined the past three decades since the original *Limits to Growth* was published; it finds no room for complacency and little more for optimism.

"Overshoot" is a term the authors use to describe the process of going beyond one's limits, often without deliberate intention. About the world's untenable resource position, the same authors conclude:

> The potential consequences of this overshoot are profoundly dangerous. The situation is unique; it confronts humanity with a variety of issues never before experienced by our species on a global scale. We lack the perspectives, the cultural norms, the habits and the institutions required to cope. And the damage will in many cases take centuries or millennia to correct ....
>
> We also believe that if a profound correction is not made soon, a crash of some sort is certain.
>
> And it will occur within the lifetimes of many who are alive today.[1]

If the authors are correct in their assumptions of approaching danger, what warning signs might help us to decide when it's time to check out the lifeboats and emergency supplies?

One essential reference is the CSIRO's *Australia: State of the Environment 2001* report, a conservative assessment of our country's natural assets.

Most of the critical indicators point to an environment in dramatic and progressive decline with few, if any, signs of improvement on the horizon.

It's not good news.

The following extracts gives you a brief overview of the national catastrophe that is rapidly developing as we, too, overshoot our limits:

> Existing pressures from human settlements are not consistent with a sustainable environment.

> Most indicators of resource consumption continue to outpace population growth.

> The volume of waste appears to have stabilised at a level which is high by international standards and there has been a recent rapid increase in the quantity of hazardous waste generated ...

> Because of lack of data on the number, location and status of contaminated sites, the environmental effects of these sites remain unknown.

> Increasing pressure to extract surface and groundwater for human use are leading to continuing deterioration of water bodies ...

> As more controls are placed on the use of surface water, more groundwater is used. The overuse of surface and groundwater resources affects aquatic ecosystems. About 26% of Australia's surface water management areas are close to or have exceeded sustainable limits.

> Water use has increased from 1985 to 1996/7 by 65% and water is overused in some regions.

Surface water quality has deteriorated further in many areas due to increasing salinity.

Dry land salinity, one of the legacies of broad acre land clearing, is predicted to affect some two million hectares of native vegetation by 2050.

The increase in salinity in the Murray–Darling Basin is causing water quality decline and land degradation. River water in several catchments is predicted to have salinity levels that will exceed drinking water guidelines within the next 20 years.

Land degradation, including erosion, is still a major contributor of turbidity, nutrients and pesticides in waterways… Since the 1960s there has been a dramatic increase in pesticide use but regular monitoring in inland waters and in groundwater is uncommon. The effects on the environment are uncertain.

… the frequency, size and persistence of harmful algal blooms in inland waters seems to have increased over the past 50 years. Algal blooms in dams cost farmers more than $30 million per year, and in rivers, storage and irrigation channels about $15 million per year.

Broad acre land clearing continues in Queensland and New South Wales. This is one of the key threatening processes of biodiversity … The rate of land clearance has accelerated, with as much cleared during the past 50 years as in the 150 years before 1945. Only four other countries exceeded the estimated rate of clearance of native vegetation in Australia in 1999 … The loss and depletion of plant species through clearance destroys the habitat and thousands of other species.

Well that's the report card.

If we'd been renting the country from indigenous landlords for the past 200 years do you think we'd get our bond back?

Tenants who leave rental properties in an appaling state of disrepair are usually presented with a bill for the cost of repairing the damage and returning the property to its original condition.

It's been estimated that the cost of repairing salinity alone is somewhere in the region of $70 billion.

White Australians should thank their lucky stars that the landlords are not in a position to collect what is rightfully owed to them.

Back to the issue at hand.

It is essential that we confront the overwhelming evidence warning us that we have long ago exceeded the sustainable limits of our land and water. As a consequence, our resource base – and in particular the quality and quantity of our available water supplies – are deteriorating rapidly. It is obvious to anyone with commonsense that to continue to grow and expand when your resources are diminishing is ultimately self-destructive.

The situation calls for a radical reassessment of our present position, not more growth. Yet growth remains the stated aim and ambition of politicians and decision makers.

During 2004, planners and water authorities in every capital city published glossy projections of future population growth and the new expanded water supplies that would service these millions of newcomers expected during the next 20 or 30 years. These "strategic plans" insult the intelligence of the reader by inviting them to share the fantasy that all of this is "sustainable".

Over and over, the documents talk about meeting the challenges posed by reduced water supplies and make extravagant promises and claims which will be impossible to substantiate in the future.

Some examples of this unfounded optimism from *A Thirst for Change – Water Proofing Adelaide 2004*, will give you the general idea.

> The Government of South Australia is committed to promoting the prosperity of the State through strategies to drive economic development and population growth. In early 2004 it unveiled a strategic plan for the future of South Australia with a goal of increasing the population by 500,000 people by 2050.

> Adelaide's water supplies will be secured well beyond 2025 through research and by monitoring technological change and innovation that **may** [my emphasis] lead to sustainable and cost-effective water sources such as desalination of seawater or brackish water.

But no amount of research or monitoring of technological change can make water where none exists. The Murray River is both diminishing and deteriorating. The only option on offer here seems to be desalination, but as we will soon discover, that is not the universal panacea for water shortages that desperate politicians fondly imagine.

The Premier of Victoria was upbeat about his state's prospects. In the foreword to *Securing Our Water Future Together*, Steve Bracks writes:

> In Victoria water is at the top of our agenda ... Victoria continues to lead the nation in sustainable water management ...

Hang on a minute Steve! When the Hamer Government began construction of the Thomson Dam back in 1976, it was supposed to "drought-proof" Melbourne for future generations, but it only bought a few extra years. As soon as Melbourne began taking water from beyond its own catchments, it lost any claim it might have had to sustainability.

Once the Thomson River fed the Latrobe Valley and the Gippsland Lakes but now up to 93% of its flow is diverted in order to fill Melbourne's storages. As a result the river often fails to flow in the lower reaches in summer. Downstream of the dam, low flows mean an excess of nutrients, increased turbidity and threats to the viability of native species.

And you call that "sustainable water management"? Steve, mate, you've got to be joking.

Undeterred, the Premier continued to outline the challenges that will, by now, be all too familiar to the discerning reader. They include the idea that growth is not only inevitable, but exciting, desirable and achievable.

There are more challenges ahead. Business, agriculture and exports are growing and need secure supplies of water. Victoria's population is growing too. By 2030, Melbourne alone will accommodate more than one million additional people and regional Victoria another 350,000. As a Government we are preparing for this population increase and growing our economy, while protecting our unique environment, recognising that a healthy water supply is pivotal to our future.

It is difficult to reconcile this confident optimism with the Minister for Water's statement on the very next page which reads as follows:

Over recent years it has become increasingly clear that our current water use is **not** [my emphasis] sustainable. Our rainfall is becoming more unreliable and, as we experience our eighth year of drought ...

Eight years of drought, unreliable rainfall, unsustainable use and even then they want to pack in another million consumers. And they still imagine that there will be plenty of water left over for "business, agriculture and exports".

The national obsession with growth and the persistent refusal to recognise the limits and constraints set by nature are a weird form of ecological suicide.

Unfortunately this collective and contagious insanity is not confined to Melbourne and Adelaide – Sydney too claims to be developing a "Plan" that, according to Premier Bob Carr, "charts our course towards a sustainable and secure water system for people and rivers over the next 25 years ... In this plan we move beyond last century's solutions."

The "Plan" is very coy about providing practical details of how exactly the NSW Government plans to move "beyond last century's solutions". However, for all its shortcomings it is a significant improvement on previous "plans" that were kept secret in order to protect the Government from public ridicule.

In fact, the Sydney planning fiasco is so disturbing it requires a chapter of its own. We'll come to that a little later on.

First, some study is needed in order to understand the three basic strategies that appear in many of the plans for the future survival of our capital cities.

Desalination, recycling and groundwater are presented as options that offer the possibility of new untapped resources that could augment our shrinking urban supplies. In order to participate in any forthcoming debate about the merits or otherwise of these alternatives, some background information is necessary.

The following three chapters examine these options in some detail.

Measurements of water

One **litre** (L) is one thousandth of a cubic metre.
A litre = 0.219 of a gallon.
A gallon = 4.546 litres.
A litre weights approximately one kilogram.

A **kilolitre** (kL) is one thousand litres, one cubic metre weighing approximately one tonne.

A **megalitre** (ML) is one million litres, one thousand cubic metres, one thousand tones.

A **gigalitre** (GL) is one thousand million (or one billion) litres of water, one million cubic metres, one million tonnes.

Desal

Great White Hope or Great White Elephant?

The State Government's proposal for a desalination plant is a confession of failure, a clear acknowledgement that it lacks an effective water strategy. It continues the sorry tradition of finding more water to waste, rather than saving the water we have. The very same profligacy, however, appears to leave us with no alternative to desalination...

The desalination project perpetuates the flawed central thinking of last year's Metropolitan Water Strategy which, despite offering many welcome initiatives, put little emphasis on re-using water.

In a report last year, the Premier's expert panel on water saw desalination as very much a last resort, if all else failed. Well, most else has failed – not least the Government's attempts to come up with an effective plan to save Sydney's water.

– Editorial, *The Sydney Morning Herald*, 30 April 2005.

"HOW WERE WE ABLE to drink up the Sea?" cries the madman in Friedrich Neitzsche's *Thus Spake Zarathustra*.

How indeed?

Desalination is the modern mantra of choice for politicians and water providers facing accusations of failing to plan for the future.

The media version being offered to the general public suggests that it's just a simple matter of taking the salt out of seawater. It's a proven technology and everybody's doing it. Why shouldn't we?

This example is extracted from the NSW Government's *Metropolitan Water Plan 2004*:

> Desalination is the process of removing salt from seawater (or brackish water) in order to produce drinking water. The reverse osmosis technique involves forcing seawater through a filtration membrane, which has very small pores that allow the water to pass but retain the salt.

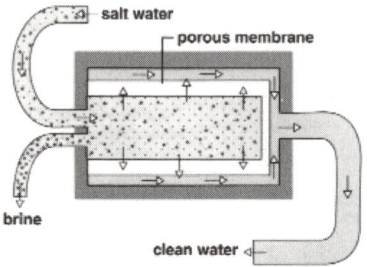

◀ *Schematic of reverse osmosis process that can generate drinking water from salty water. (If only it were this simple! J.A.)*

> Desalination plants are used in many cities around the world to provide useable water. They do provide a reliable supply and good quality water and are immune from drought and climate change impacts. While they are currently more expensive and energy intensive than traditional water supply options, research and development is reducing the costs and improving their energy efficiency.[1]

So that's it. We put in a series of massive desalination plans and our problems are solved. After all there's plenty of water out there in the ocean.

Now we can relax and forget about the so-called water crisis. Let's raise our glasses of water – desalinated of course – and drink a toast to good old desal. Science and technology have finally come to the rescue and now we can leave all the fine details to the army of planners, consultants and engineers waiting patiently in the wings.

This probably sums up the average citizen's initial response to desal – until they learn a little more about the process. However, before we explore the mysteries of desalination, ask yourself this question: If it's such an obvious solution, why on earth didn't somebody suggest it before? Why have we waited so long? After all, the technology has been around for 60 years at least.

Sixty years of application should have finetuned the process. Look at cars – how much they've changed in sixty years – or refrigerators or railways. We've come a long way since then, but unlike these straightforward mechanical inventions, the history of the desalination of seawater is not one of continuous progress, but of optimism followed by failure and disillusionment; these cycles are still being repeated around the world today.

Take the example of the United States, where reverse osmosis was perfected by the Indian scientist Sourirajan at the University of California, in the early 1950s. The process was considered to be such a massive breakthrough that the US Department of the Interior put up $30 million (a lot of money in the 1950s) specifically to develop the desalination of seawater to overcome water shortages in California and elsewhere.

By the 1960s US senate reports indicated that the project was well under way. A 1962 report was positively upbeat:

> Once the salt can be removed from the oceans around us, the nation's water problems will be well along to solution.
>
> Industries and agriculture in some sections will come practically to a standstill unless the sea can be made to give up its precious moisture ... A few areas of the country already face a semi-critical water shortage ... Scientists and engineers are working at almost an emergency pace to develop the plants and processes that will be needed to do the job.
>
> At least half a dozen different methods are under study to convert salt and brackish water into fresh water. Some involve distillation, others freezing, others electrodialysis or passing water through a membrane to filter out the salt ...
>
> One or a combination of these processes will be adopted as the fastest and most economical way of taking fresh water from the sea. Within a decade, mammoth sized plants will be functioning ...[2]

The US senate was treated to a glowing vision of a vast network of desalination plants pumping freshly made water into giant canals flowing inland, turning the desert wastes into productive farmlands.

US President John F. Kennedy personally supported the project. In one of his oft-quoted addresses to the nation he said: "If we could ever competitively; at a cheap rate, get fresh water from saltwater, [this] would be in the long range interests of humanity which would really dwarf any other scientific accomplishment."[3]

But in spite of the millions of dollars invested – and all the time and energy and expertise – within a decade the massive US desalination project was gradually wound down and phased out. The pilot plants were abandoned and without any public explanation, desalination quietly vanished from the national agenda until the 1990s.

Why?

Because in the final analysis they discovered that reverse osmosis on a large scale was too complicated, too unpredictable and too expensive to produce large quantities of drinking-quality water (or even lower-grade water for agriculture) from the sea. The entire project was an embarrassing and costly failure that no one in the industry wants to remember.

The only beneficiaries were the corporations, scientists, consultants and engineers who developed and worked on this giant white elephant that was supposed to drought-proof North America.

By the way, the US is having another serious drought, and desalination still doesn't seem to be making much of a comeback.

However, here across the wide Pacific, a "new" and "improved" version of the same old elephant is being paraded by another enthusiastic group of corporations, consultants, scientists and engineers as an ideal strategy to drought-proof our cities.

The desal sales brigade have run a very effective media campaign in Australia. One by one, visiting international experts pointed out – without citing actual examples – that the cost of desalination has come down; they extolled the value of "economies of scale", and talked up the advantages of the "new technology".

The Australian newspaper's environment writer Bernard Land, for example, was positive and enthusiastic on 17 November 2004:

> It's getting cheaper to make seawater drinkable, and more expensive to get water the traditional way from the dam to the tap.
>
> "Taking the salt out of water for drinking remained slightly more expensive than conventional water treatment methods but the gap was closing", visiting international expert on desalination Tom Pankratz said in Sydney yesterday.
>
> "The technology is established to the point that people are no longer afraid to build a relatively large plant and the economies of scale really kick in," said Mr Pankratz.

With due respect to his expertise, Mr Pankratz is hardly an objective expert. He is vice-president of the giant US-based consultancy CH2M HILL which has been actively promoting both desalination and it's own consultancy expertise around the world for years.

Mr Pankratz went on to suggest that "the only real new source is seawater and desalination – so it should be part of any long term strategy."

If this "new" and "improved" US technology is in fact the best solution to our water crisis, why aren't *they* using it to drought-proof some of their large cities currently affected by severe water shortfalls?

After all, they have the benefit of all of those 60 years of experience behind them.

Perhaps we could learn why by examining the only large functioning state-of-the-art seawater desal plant in the United States at Tampa Bay in Florida. Optimism was certainly in the air when work began on the project in 1999. Promoted by Tampa Bay Water as "the largest desalination plant in the western hemisphere", its supporters boasted that it "would produce the cheapest desalinated water in the world".

Costing US$110 million, the plant's projected output was 125 mega-litres per day – equivalent to 2 hours of Sydney's daily consumption. However, there were so many unforeseen problems and delays that the public began to lose faith in the project.

Finally, four years later in March 2003, a test run generated the following public statement:

> "The facility had failed key operational tests, notably concerning pre-treatment removal of impurities before actual desalination … Asian green mussels clustered around the intake pipes … were clogging up the system, fouling the expensive membranes used for removing the salts and thus reducing the membrane's anticipated five-to-seven year life.[4]

This was the last straw for the stressed construction company, Covanta Tampa Construction, which declared bankruptcy and abandoned the project.

Then other faults began to appear. According to a spokesman for Tampa Bay Water in July 2004: "The plant design is flawed, the intake system is lacking and pre-treatment system is inadequate to protect the membranes," (*Global Water Report*, Issue 201, 20 August, 2004.)

It is estimated that it will cost an extra $50 million to fix these and other problems.

Waterweek, the journal of The American Water Works Association lamented in its August issue that

> Although the $110 million, 125-megalitres per day plant was finished in April 2003, it has not functioned properly from the beginning stemming from flaws that caused cartridge filters to clog in the pre-treatment phase of the operation. While the plant now produces drinking water from ocean water at a rate of about 12 mgd, it does so in a costly, inefficient and intermittent way, according to Tampa Bay Water.

The plant is expected to be fully functioning in 2006, **seven years after it was first commenced**.

Bear in mind that this is the most up to date seawater desal plant in the whole USA. Not exactly a shining example is it?

Before the ill-fated Tampa project, only two American cities had invested in full-blown seawater desalination plants – Key West, Florida, in the early 1980s and Santa Barbara, California, a decade later. Both of these cities shut down their operations shortly after they began in favour of cheaper options, but the facilities are still maintained in case of emergencies.

So the Tampa plant was a big breakthrough for the desalination industry, a chance to prove that now, after all these years, the problems and failures of those earlier plants had been solved.

"Desalination has never had a lot of public support in the United States," Fared Salem, DuPont's desalination manager told author Jeffrey Rothfeder back in 2000. "But maybe when people see how well Tampa works and how limited the financial pain is, there will be renewed interest in a technology that can really address the twenty first century's water problems."[5]

Instead of renewing interest in desalination, the Tampa Bay fiasco was enough to make other prospective desal clients in the US nervous. The San Diego County Water Authority abandoned its plans for a $270 million plant on the Pacific coast, and several other US customers also put their plans on hold.

All this was happening in July and August 2004, about the same time that Frank Sartor, the NSW Utilities Minister, announced that desalination would help to guarantee Sydney's future water supply.

"Desal, I think, in the long term, is probably inevitable, especially if we can get energy costs down, because we do not want to create a greenhouse problem for ourselves," Sartor told a NSW Estimates Committee hearing.

One of the most important issues concerning desalination (when and if it *does* work) is the final cost of the finished product. Estimates vary from around 80 cents to $3 per kilolitre. An "Economic and Technical Assessment of Desalination" commissioned by the Commonwealth

Department of Agriculture in September 2002 put the average cost of desalinated water at $2.20 per kilolitre. (The full study can be downloaded – see notes)[6]

Power costs are estimated in kilowatt hours per kilolitre (kwh/kL). While some industry lobbyists are suggesting that desal power consumption has dropped from 22 kWh/kL to 2 kWh/kL, a typical RO plant similar to that in Perth consumes about 6 kWh/kL.

To give you some idea of what that means in practice: a series of desalination plants capable of producing Sydney's daily consumption of 1,676 megalitres would consume around 10 million kilowatt hours. (1,676 megalitres = 1,676,000 kilolitres x 6 kWh/kL = 10,056,000 kilowatt hours of electricity for one day's supply).

This a vast level of energy consumption.

Most politicians who have become overnight experts on desalination are fond of telling the media how the cost of desalination is falling rapidly.

If this is the case then it is hard to explain a September 2004 debate in a US congressional subcommittee over legislation to "authorise the Department of Energy to spend $200 million over 11 years to subsidize the high electricity costs faced by desalination plants, such as the one operated by Tampa Bay Water in Florida". (*Waterweek*, 15 September, 2004)

The subcommittee was also told that "many cities across the United States are facing impossible price tags to maintain their water systems, including finding money to meet high energy costs" – and these are cities *without* desalination.

Energy costs are only half the story. Energy *supply* is the big issue. Energy supply in NSW is already stretched to the max. When Sydney experiences a very hot day and all of those domestic air-conditioners purr into action, the integrity of the entire power grid is threatened.

The additional power output necessary for a large desalination plant are both impressive and disturbing.

The current Sydney proposal to build a facility capable of producing 100 megalitres of water a day – supplying a mere $1^1/_2$ hours of Sydney's daily needs – will require enough energy to produce 255,500 tonnes of greenhouse gases every year.

Maybe that's one reason why many of Perth's citizens were less than enthusiastic about the prospect of desalination.

If only half of the desalination plants under consideration by desperate councils and water authorities across this country proceed as planned, the drain on the nation's power grid will be considerable.

At least five are currently proposed for NSW coastal communities suffering from severe water shortages.

"Desalination is the only answer," says Brendan Pavier, the Mayor of the fast growing centre of Wyong 100 km north of Sydney. Dams on the surrounding Central Coast are down to 24% of capacity and falling. Many other drought affected communities are examining their options and coming up with similar answers.

But even if the plants do work as planned, where will all the power come from? And what about the greenhouse gases, and the likelihood that a radical increase in emissions will in turn accelerate climate change which will prolong the drought?

It's an unpleasant loop with few practical or realistic answers. The NSW Government proposes to reduce greenhouse gases by:

> Using low-emission or renewable energy sources; using waste heat from nearby industries; gaining synergies by combining a desalination plant with a co-generation plant; and using offsets (such as tree planting, as is proposed in Perth) linked to the Government's existing Greenhouse Gas Abatement Scheme.[7]

None of these strategies seem to work very effectively elsewhere. Tree planting, for instance, is unlikely to compensate for the large volume of pollution, and the idea of planting thousands of acres of trees in a time of drought, given the volume of water required to sustain them, could hardly be called practical.

Perth's planners are examining the possibility of using windpower to generate the 24 megawatts (a megawatt is a million watts) of power required annually but the actual site for the giant turbine generators had not been finalised at the time of writing.

There is another possibility, but at this stage it would be political suicide to endorse it.

On the horizon of the desalination debate is the dark shadow of nuclear power. Strong arguments are emerging internationally that nuclear power is the answer to the high-energy inputs necessary for large-scale desalination.

There are already several pilot projects that harness nuclear power for this purpose, the most recent being in Chennai, India. Other countries are keen to follow their example.

Egypt's Prime Minister, Ahmed Nazif, assured Western diplomats in January 2005 that the principal function of the country's nuclear program was to provide enough power for desalination of drinking water.

It is no secret that 'Nuke desal' is a hot topic for debate among Australian power brokers, even though the proposition has been publicly rejected by several state premiers and the Federal Government at this point in time.

But times change, and so do attitudes.

For example, the NSW Premier Bob Carr, was originally vehemently opposed to desalination, contemptuously referring to it as "bottled electricity" because of the high levels of greenhouse gases involved. However, after prolonged exposure to the persuasive arguments of the friendly corporate door-to-door desal salesmen, Mr Carr announced that he had finally seen the light. The lucky winners of the government's $4 million desalination feasibility study tender – Gutteridge, Haskins and Davey (GHD) – celebrated with champagne, anticipating, as one insider observed, that the best is yet to come.

The "feasibility study" is not about whether Sydney will desalinate, since this seems to be a foregone conclusion – it is about how, when and where. The group will work with Sydney Water to develop attractive proposals to present to a community that has not yet been consulted about their attitude to desalination, nor have the power implications been made clear to the public. It is no secret that the possibility of nuclear power has been often suggested as a "green" option to beef up Sydney's ever-increasing energy needs, while at the same time reducing controversial greenhouse gas emissions.

With or without nuclear power, desalination is not a popular choice once it has been fully explained.

A survey of Perth residents, published in *The West Australian* in November 2004, showed that only 25% believed that the WA government should proceed with desalination. Seventy-five per cent of people were opposed. If the nuclear option were included, this number would probably rise considerably.

The "green" pro-nuclear argument goes like this:

> Using nuclear power to desalinate seawater would mean that tree-planting programs (used to offset greenhouse gases) could be seen to actually reduce net carbon dioxide in the atmosphere. And the more seawater we use to provide water for our coast-hugging population, the more rainwater we could leave in our rivers to keep them environmentally healthy. (*The Canberra Times*, 22 September, 2004)

Nuke desal for a healthier environment – I wonder if the electorate will go for it.

Optimum benefits of the combination can best be achieved if the facilities are located close to one another, preferably side by side. Since desalination plants need to be near to their water source, in the case of Sydney this would mean a beachfront site somewhere within the city's environs. Tamarama, Bondi, Coogee, Manly and Palm Beach would all be excellent locations, although I suspect the proposition would meet with some resistance from die-hard residents reluctant to embrace the new technology. So in the final analysis it will probably be sited somewhere out of the way, somewhere like

La Perouse where it could be combined with the other heritage tourist attractions in the area.

Desalination plants are not very attractive buildings. The most flattering description the Western Australian Government could come up with was: "It will look like a typical Bunnings warehouse." Nice! But of course we can always design one in a neo-Tuscan style and paint it in those wonderful natural heritage colours so that it will blend in with its surroundings.

The other issues to bear in mind if the authorities are considering building a desal plant in your area is the noise and the pollution.

The official industry training manual is the *Water Desalting Planning Guide for Water Utilities*, published in June 2004. Its authors offer water authorities the following helpful advice: "A project located in an environmentally sensitive area or in a sensitive neighbourhood will require more effort to implement."[8]

Why might this be?

The *Planning Guide* identifies the major controversial environmental issues as: noise emissions, air quality and odours, and marine pollution.

Reverse osmosis desalination processes generate high levels of noise emissions from all the pumps, valves, air blowers and flow-restriction devices. Theoretically, these can be minimised by designing the building appropriately.

Under the heading of "Air quality and odour", the planning guide identifies the unpleasant odour emissions associated with the removal of gases contained in seawater:

> Hydrogen sulphide (known as rotten egg gas) is the most common of these gases and can lead to extreme problems if not properly treated. In addition to having an extremely obnoxious odour, this gas is also toxic and significantly corrosive to electrical and electronic components. The threshold for detection by human beings is approximately 5 parts per billion ....[9]

Desal plants can be noisy, smelly neighbours, but it's their long-term impact on the marine environment that should concern us most.

Desalination is often described as a technology that separates the salt from seawater. This sounds simple in theory but the actual practice is far more complex.

> The most widely used process for seawater desalination uses reverse osmosis ... requiring 100,000 pounds per square inch of pressure as seawater is pumped into plastic membranes through which salt and other molecules cannot pass. The reverse osmosis membranes today cost half as much, last twice as long and are twice as productive as membranes manufactured ten years ago.[10]

All desalination plants generate a lethal by-product – a poisonous concentrated saltwater mixed with toxic chemicals. Chemicals are essential in order to prevent salt corrosion and the build-up of scale in the plants. Acid is added to the incoming seawater to reduce its alkalinity and prevent the formation of calcium carbonate. Highly toxic scale-inhibiting chemicals are also used to protect the pre-filters.

For every litre of desalted water, a litre of poison is pumped back into the sea.

In the case of the 100 megalitre/day plant proposed for Sydney, for example, this would result in 36,500,000,000 litres of waste each year. If you combine that with the one billion litres of sewage pumped into the ocean through the city's deep ocean outfalls every day, 365 days a year, you come up with a total pollution level of 400 billion litres of toxic waste pouring into the sea annually – almost enough to fill Sydney Harbour.

And that's just for one city!

Among the OECD countries, Australia ranks second only to the US in its production of waste. Widespread desalination would add considerably to this burden.

World wide, more than 13,000 desalination plants produce around 20 billion litres of waste daily. And many more are proposed. We have already stretched to the limit the ocean's capacity to absorb the vast quantities of waste pouring out of the world's cities, and now we are getting ready to add more.

And the crazy thing is that by doing so we are poisoning the very source we want to desalinate. It's a slow form of ecological suicide but it appears to be gathering momentum.

The composition of life in the oceans has already changed radically in response to our poisoning. Most waste eventually ends up in the sea. The scientific theory, if you could call it that, is one of infinite dilution. The idea is that the ocean is so vast that it doesn't matter how much toxic material or excrement or radioactive waste we dump into the big blue.

This is a big mistake.

Scientists once reassured us that it was impossible to irreversibly damage the atmosphere of the planet, but then, whoops, sorry, they now tell us that they made a serious and perhaps fatal miscalculation. One that provides yet another illustration of a serious flaw in the theory of infinite dilution.

Take the example of Sydney's infamous deep ocean outfalls.

In his book *Planetary Overload* (1993), Professor A.J. McMichael offers the following insight into Sydney Water's environmental philosophy:

> Recently, I was consulted by the waste-water authority in an eastern state of Australia about their plans for reducing coastal pollution (including the contamination of edible fish) by heavy metals and organochlorine chemicals from industrial effluent. They proposed a deep-water offshore outlet, and explained that fortunately it was "a high-energy coastline" – meaning that wastes would quickly be flushed out to the deep ocean where, they said, "it is no longer our problem".[11]

When the outfalls began to operate in 1991, it was found that the experts were wrong. The sewage did not magically disappear into the unfathomable deep. Instead, as Ann Young explained, "the sewage seems to have dispersed only across the inner shelf (up to 10 km off-shore) and to move parallel to the coast rather than moving right away. A chemical produced in the digestive tracts of animals, coporostanal, was traced as an indicator of sewage and was found accumulating in the sediments of the inner shelf." [12]

So ultimately Sydney will end up desalinating its own diluted toxic waste, and then dumping the concentrated residue back into the sea again, and again and again and again.

As Macbeth so aptly put it:

> This even handed justice
> Commends the ingredients
> Of our poisoned chalice
> To our own lips.

Thanks to the uncounted billions of tonnes of chemicals that have found their way to the sea via run-off or sewage or at-sea dumping, the oceans of the world are indeed a poisoned chalice, many of whose most dilute ingredients will not be removed by reverse osmosis desalination.

PCBs (polychlorinated biphenyls) are typical of the modern poisons that can be found in all of the world's oceans. PCBs were commonly used to dissipate heat in electrical transformers as well as in hydraulic fluid, lubricants, fire retardants, paints, varnishes, inks, carbonless copy paper and pesticides. They were, according to the scientists who invented them, perfectly safe.

For 40 years users of these chemicals dumped them in landfills, along roads, into sewers and water bodies without thinking of the environmental consequences. Then in a landmark study in 1966, designed to detect DDT in the environment, Danish researcher Sören Jensen reported that in addition to DDT he had found PCBs to be widespread as well. [13]

Since then PCBs have been discovered in marine ecosystems everywhere. They are found in the highest concentrations in carnivorous fish, seabirds and mammals, human fat, and human breast milk. The dilute oceans have carried the poisons of the industrial world into even the most remote areas of Antarctica and the Arctic Circle.

> DDT and PCBs are the only organochlorines that have been monitored on a systematic basis in Arctic marine mammals … The PCB levels in the breast milk of Inuit women are among the highest ever reported …

> These results suggest that toxic compounds such as PCBs could play a role in the impairment of immunity and the high occurrence of infection among Inuit children.[14]

PCBs are just one of thousands of different compounds with which we have effectively poisoned the oceans, the tiny plankton and the fish and mammals that form part of the marine food chain that humans ultimately consume.

These chemicals affect humans at levels so dilute that they are not detectable in sea or fresh water. Can we trust desalination to remove them?

Yes? No? Maybe?

Only time will tell.

The oceans of the world – the raw material for this vast experiment – are the last great frontier. We know so little about them, in spite of our arrogant assumptions of scientific insight, that we run the risk of forgetting that we live in a global environment that is constantly changing in response to the inputs we are providing. These changes can be sudden, unpredictable and often irreversible.

You can't simply turn around and say "sorry" to nature after the damage is done – that's not how the system works.

Consider, for example, the increasing acidification of the oceans caused by excess atmospheric carbon dioxide. A British climate change conference in Exeter in February 2005 was told that even slight changes in the pH of the oceans could wipe out coral reefs, many species of fish and other marine life.

Carbon dioxide is absorbed by the sea in vast quantities. Of the estimated 800 billion tonnes of CO_2 generated by humans since the beginning of the industrial revolution, more than 50% combined with sea water. Carbon dioxide produces carbonic acid when mixed with water and this is progressively making the oceans more acidic.

"Scientists did not look at this problem because everyone assumed that the chemical composition of the sea was constant. But this change is elementary chemistry and we missed it," said Dr Carol Turley, Head of Science at the Plymouth Marine Laboratory.

Much of the excess carbon is fixed in the shells of trillions of tiny plankton which live on the surface of the ocean. When they die, their shells sink to the seabed, taking the carbon with them. These creatures cannot survive in an increasingly acidic environment. As they disappear, the removal of carbon from the atmosphere will gradually cease, and greenhouse gases will multiply exponentially.

"These creatures are part of our survival bubble," Dr Turley warned. "They're the lungs of the planet. People have not woken up to the potential impact their removal will cause ... Many of the species we rely on to eat, like cod, will disappear. The whole composition of life in the oceans will change." (*The Sydney Morning Herald*, 5-6 February, 2005.)

The oceans are no one's property, they are not divisible, their perpetual motion means that they wash every shore no matter how remote. Although their depths are still mostly unexplored, the capacity of the seas to moderate and modify the earth's climate is universally recognised.

But it is not for this alone that we sit by the waves in times of deep distress or personal crisis. There is another deeper spiritual dimension to our relationship that is beyond our understanding. This is as close as we can get to the mother of all things, to the compassionate, creative, eternal energy of the universe.

In 400 BC, the anonymous author of the *Ashtavakra Gita* wrote:

> I am the boundless ocean
> This way and that,
> The wind blowing where it will
> Drives the ship of the world,
> But I am not shaken.
> I am the unbounded deep
> In whom the waves of all the worlds
> Naturally rise and fall,
> But I do not rise or fall.
> I am the infinite deep
> In whom all of the worlds
> Appear to rise.
> Beyond all form,
> Forever still
> Even so am I.

It is time we listened to the subtle voices of nature instead of those of the fanatics who simply see the world as a resource to be exploited in order that the Gods of Growth be appeased.

Desalination of the sea is not the answer to our water problems. It is survival technology, a life support system, an admission of the extent of our failure.

Let's face it, when we can no longer drink the water from our creeks and rivers, when our own actions finally cause the rains to fail and our taps to run dry, then we must face the fact that sustainability is moving beyond our grasp, that it's a fantasy we can no longer believe in.

We will have reached a critical point in our progress towards extinction, the line in the sand that marks the end of living and the beginning of survival.

From Toilet to Tap
The NEWater cycle

The use of recycled water to augment drinking water and for other purposes is expanding throughout the world ...

> – NEWater Promotion, Singapore 2003

Greater use of recycling can help resolve our water resource problems, but the community must continue to have trust and faith in its water authorities. Maintenance of safety standards and the health of consumers is paramount. One failure, and all investment in recycling will have been put at risk.

> – John Radcliffe, *Australian Science*, July 2004

WATER RECYCLING IS HIGH on the agenda of water managers and their spin doctors in the Twenty Thirst Century. They assert that water recycling is essential and that it makes sense both financially and environmentally. After all, only 2% of our daily water consumption is actually used for cooking and drinking, so why not substitute treated recycled water for use in our gardens or for flushing toilets? But do we want to use recycled water for bathing the baby or washing our clothes or cleaning the floors?

Do we want to drink it?

The main problem with recycling water for domestic use is the cost of the plumbing. If houses are to have a dual supply – one pipe for drinking water and one for recycled water – then it is clearly not a viable proposition to duplicate the thousands of water mains that already service our cities. The work would take years to accomplish and would cost uncounted billions of dollars.

So that's clearly not going to happen.

Another alternative is to pipe recycled water to individual industries, farms and businesses that are high water users. There are many such schemes, but at present their overall impact on national water consumption and reuse is negligible.

The final option and the only one that looks practical in engineering and economic terms is to pump the recycled effluent back into the drinking water supply, either into the dams of cities and towns, or at some point in the mains pipelines.

That's where the discussion about water recycling begins to get interesting. While there's no argument with the concept in principle, opinions vary widely about how it should actually work in practice. What water should be recycled, for what purpose, by whom and for whom? These four basic questions need to be answered if recycled water is ever to become a significant part of the nation's water supply.

When politicians talk about recycling water, they are usually talking about big schemes that sound exciting and innovative when they're announced to the media. But the failure of many of these high-profile recycling projects seldom receives the same media attention.

Three examples follow.

Back in 2000 with much fanfare, the NSW Government presented the electorate with *Waterplan 21*, its development blueprint for achieving "sustainable water consumption in Sydney by 2021". An important aspect of this plan was the construction of a pipeline to supply up to 50 million litres of treated recycled sewage effluent daily to industrial customers in the city's southwest.

Sydney currently recycles only 30 million litres out of its total daily output of 1800 million litres, about 1.5%, so this new plan was promoted as a substantial step in the right direction.

After four wasted years of work and planning costing around $20 million, the project was suddenly scrapped in June 2004, not because it was unrealistic but apparently because not enough industrial

RECYCLING TERMINOLOGY

POTABLE WATER: Water deemed suitable for human consumption without deleterious health risks.

NON-POTABLE WATER: Water that does not meet drinking water guidelines.

EFFLUENT: A liquid extracted from sludge at sewage treatment plants. When refined to a high degree, effluent may resemble water, but its precise chemical composition is not known. Its long-term effects on humans and animals are the subject of international scientific debate and controversy.

WATER RECYCLING/REUSE, WATER RECLAMATION: Misleading terms used by the water industry to foster the illusion that effluent is the same as water. The California Water Code gives the following definition – "water which, as a result of treatment of waste, is suitable for a direct beneficial use or a controlled use that would not otherwise occur".

DIRECT POTABLE REUSE: The addition of treated sewage effluent to drinking water supplies.

INDIRECT POTABLE REUSE: The release of treated effluent into groundwater, rivers, streams and dams which ultimately become drinking water supplies.

UNPLANNED OR INCIDENTAL INDIRECT POTABLE REUSE: *Water Recycling In Australia* uses this intriguing term to describe cities and towns who extract their water from effluent-contaminated rivers. The book cites New Orleans, London, Kyoto and Adelaide as examples.

DUAL RETICULATION: The provision of a second mains system designed to provide a separate recycled water supply from the treatment plant to the consumer.

customers could be found. That was one explanation. However, an industry newsletter, *Platts Global Water Report* (9 July, 2004) put forward a more likely reason for the cancellation:

> Insufficient demand was blamed for the decision; however,
> public officials said it largely resulted from funding cutbacks.
> The state government is running a budget deficit and stands
> to save $90 million from not proceeding with the pipeline.

So you can see from this failed project that water recycling is not as simple as it appears. There are substantial costs to the taxpayer in order to subsidise the supply of lower-cost water recycled to industrial users who, one would imagine, would be keen to maximise their profits at public expense.

And if not, why not, one wonders?

If industrial consumers reject recycled water as unsuitable for their purposes, then what is it good for? And who will use it?

South Boulder is an isolated mining town in one of the driest parts of Western Australia. When the town council opened its modern recycling plant in 2002, it thought customers would be queuing up at the door. But a year later the town had not found a market for the surplus effluent and it was going to waste.

Finding new customers was not a problem in Victoria, where another large innovative recycling scheme seems to have hit another brick wall – that of government inertia.

The up-market Sandhurst Club is a large residential development (1850 sites) built around two golf courses at Carrum Downs on the outskirts of Melbourne. In February 2003, as part of National Water Week, Victoria's Environment and Water Minister, John Thwaites, officially launched Sandhurst as the state's first residential recycled water project. A million buckets of Class A recycled water would be made available for household use every day, making Sandhurst the showcase of the government's "sustainable water use policy". A high-level task force from Melbourne Water, the EPA, the Health Department, and the Environment and Sustainability Department was established specifically to ensure the success of the model Sandhurst project. The media lapped it up. The government's environmental credentials soared.

But then, something went wrong.

Eight months later, an angry Ken Roache, Chairman of the Sandhurst Joint Venture, called a press conference to air his grievances:

> Essentially it is a situation where the government has made promises and entered into agreements whereby we've spent well over $1 million on running recycled water to every household. We were promised A grade recycled water by September. At this stage, they're saying we may not be getting it. (*Herald Sun*, 24 December, 2004)

In February 2005, the Sandhurst Club had still not received its Class A water. A disillusioned Ken Roache told journalists that these delays cast doubts on the government's ability to deliver on the new technology and its own environmental policies.

It seems that there may still be a few problems with the "new technology".

The advanced membrane filtration processes used to remove the more complex contaminants from effluent are similar to those used for desalination, and have the same drawbacks. They are prone to fouling and damage, expensive to replace and highly energy intensive.

The increased use of advanced oxidation and membrane filtration will lead to a considerable rise in greenhouse gas emissions. Typical estimates for Melbourne, for example, indicate that if the city's 20% recycling target could be met, it would result in the production of 28,000 tonnes of additional carbon dioxide. The additional energy required to transport recycled water over long distances would also add more greenhouse gases.

Over in Singapore, recycled water is being bottled for human consumption as part of a clever marketing strategy to encourage the population to "embrace" recycling.

Professional spin doctors seem fond of the word "embrace" in conjunction with new (and often experimental) technologies such as recycling and desalination. The satirist Ambrose Bierce once described "experience" as "the wisdom that enables us to recognise as an undesirable old acquaintance the folly that we have already embraced".

To continue on the subject of follies already embraced. Proponents of the Singapore experiment, called the NEWater project, claim that their recycled product is superior to ordinary tap water:

> NEWater is so reliably pure that it can be added to existing drinking water reservoirs as a supplement to Singapore's raw water supplies. This concept is supported by worldwide experience and the conclusion of trustworthy experts.[1]

More often than not, these "trustworthy experts" turn out to be consultants or employees of the companies involved in the enterprise. The "worldwide experience" mentioned in the press release is limited to a couple of experimental projects in remote places like Namibia, for instance, rather than huge cities such as London, Paris or New York.

Nevertheless, on the advice of their unnamed, "trustworthy" experts, Singaporeans are now drinking their own treated effluent with gusto, according to Ian Law, an Australian consultant to the NEWater project. Mr Law admitted that initially many people were disturbed at the thought, but that an intensive "public education" campaign had turned public opinion around. Apparently the politicians were especially convinced. "The whole cabinet was quaffing bottles of NEWater," Law told *The Sydney Morning Herald*'s Richard Macey (20 October, 2004), "I would be very happy to drink it."

One person who was not so happy to drink NEWater was the NSW Premier, Bob Carr, when offered the opportunity in Parliament in October 2004. Perhaps he was just being wary about cheap imports. Maybe he wouldn't be so choosy if it was a local brand.

In September 2004, hundreds of residents of the Sydney suburb of Glenwood were treated to a preview tasting of the local brew, after an accidental cross-connection pumped thousands of litres of recycled sewage effluent into their daily drinking supply. The duel-reticulation scheme was implemented in the early 1990s in new housing estates in the Rouse Hill area as a model for the future. Sewage effluent is treated before being delivered through a separate pipeline back to homes where it is used for gardens and flushing toilets.

Residents in the 82 affected houses awoke one day to find a letter from Sydney Water under their doors warning them that their tap water could taste salty or smell like the recycled supply. Perhaps it says something about the quality of Sydney's drinking water that nobody had previously complained.

Although NSW Health experts advised that, in their opinion, "the water was still safe to drink", free bottled water was supplied to those who doubted, until the problem was finally rectified. Sydney Water's Managing Director, David Evans, however, did not want to reassure anyone that the water was safe. Instead he released a statement saying that "the chances of customers becoming ill was extremely low". (*Blacktown Advocate*, 1 September, 2004.)

The public's trust in the quality and integrity of drinking water and those who supply it is a critical issue if domestic water recycling is ever to be implemented on a large scale.

And there's often a big gulf between the rhetoric of the mission statements and the reality of day-to-day performance.

In the case of Sydney Water, the statement reads like this:

> Sydney Water, in its management of the community's daily drinking water supplies, holds a great deal of public trust. They trust us to do the job right so they can safely drink our water, and they trust us to fix any problems quickly. How we deal with such incidents – both major and minor – will reflect in the level of trust and respect we receive from our customers, the community and stakeholders. If we do the job right, then the level of trust and support will remain high. Do it poorly and we will lose that trust very quickly...[2]

In January 2005, *The Daily Telegraph* obtained documents under the Freedom of Information Act that showed there had been at least three other such incidents in Sydney since 2001, and it appeared that residents had not been adequately informed.

Several residents experienced stomach problems and skin rashes. In response, a local GP advised one affected family to take a course of worm tablets after they all complained of various infections. "My little daughter started having warts on her face, lips and hands," one woman told *The Daily Telegraph* (12 January, 2005). "As soon as we moved in [February 2002] we started visiting the doctor a lot – we were in perfect health before so it was very unusual for us … We were very shocked when we found out about the water, we started talking about selling the house and moving somewhere else."

In February 2005, the NSW government announced that a public–private partnership would supply 150,000 new homes in Sydney's ever-expanding western suburbs with 80 billion litres of fresh recycled effluent every year. The cost was estimated at $560 million. The potential cost to the health of the residents was not included in that estimate.

We'll see the end results in 20 years or so.

An opinion poll carried out in 2003 indicated that 99% of the Australian community supported the use of recycled water for lawns and gardens. This dropped down to 49% who said they would use it for washing clothes.

Only one per cent were in favour of adding it to the drinking water supply.

Associate Professor Greg Leslie, from the UNESCO Centre for Membrane Science and Technology, was an expert consultant to the Singapore NEWater project that used a membrane technology he developed 20 years ago. He says that the public needs to be convinced that recycled water can be a superior product.

A Sydney Water spokesperson once remarked that water reuse would take off if only "the public was made comfortable with the idea that it was safe". Whether it *is* in fact safe often seems less important than the public's attitude, which was perceived to be the real impediment.

A 2003 CSIRO Literature Review of *Factors Influencing Public Perceptions of Water Reuse* explored in depth the issue of public concern "about the safety of using recycled water, considering the potential lethality of pathogens in the water and the unknown impact of chemicals used to treat the water".

The authors raised the issue of "disgust", which they defined as "the emotional discomfort generated from close contact with certain unpleasant stimuli" such as "excrement, urine, saliva, dirt and mud", and then speculate why this might lead to "avoidant behaviour".

One interesting explanation for this irrational response was the law of contagion. For those ignorant of the intricacies of the scientific mind, the document explains that:

> This law suggests that a neutral object may acquire disgusting properties from another object through brief contact (eg hair in the soup). So regardless of the fact that recycled water has been treated to highest standards [sic], people may still perceive the water to be disgusting because it has been in contact with disgusting stimuli, in this case human wastes. Research in the past has successfully measured people's disgust sensitivity to predict the probability of avoidance in the presence of unpleasant stimuli.[3]

Instead of conceding that this may be a logical and normal human reaction, these experts continued to wonder why some people are so resistant to persuasion, while explaining the difference between public and "expert" perceptions.

> Why do the public perceive such a risk in using recycled water despite constant assurances from experts and the authorities that the water is more highly treated than scheme water? From the general risk literature, it is evident that risk perceptions are different between the experts and lay people. The public tends to capture a broader conception of risk, incorporating attributes such as uncertainty, dread, catastrophic potential, controllability, and equity into their risk equation. The experts, on the other hand, define risk in terms of event probabilities and treat subjective factors as "accidental" dimensions of risk. These "accidental" dimensions of risk may however play an important role in forming people's attitudes

towards a risky situation. For example, the experts may consider that a one in a million risk of getting sick from drinking recycled water was acceptable whilst this risk may be completely unacceptable to the public, as that one case could be them or their child.[4]

In the end, the experts contend, it all comes down to an acceptable level of risk – and who decides whether the risk is acceptable?

The experts of course.

However, not all experts agree with the proposition that recycled water is "safe" to drink. Some believe that the widespread use of sewage effluent presents unacceptable risks both to the public and to the wider aquatic environment.

The arguments for this view are both disturbing and convincing. They focus on the unknown actions and interactions of the vast array of chemical and pharmaceutical compounds that end up in sewage.

While many scientists and engineers are keen to expedite the process of recycling, health authorities and the general public remain less than enthusiastic.

And with good reason.

Sewage treatment plants separate the liquid from the solids. The liquid is called effluent, the remainder is referred to as sludge. In Australia, this toxic sludge, aggressively marketed as "bio-solids" by water authorities, is spread liberally over drinking water catchments, fields and grazing pastures. Floods eventually carry some of these residues back into rivers, creeks and dams where they accumulate in the sediment.

Why is this practice banned in other countries?

Because they do not consider the risk to be acceptable.

Why not?

Treated sewage sludge contains up to 60,000 toxic substances and chemical compounds, including:

- polychlorinated biphenyls (PCBs)
- cholorinated pesticides – DDT, dieldrin, Aldrin, endrin, chlordane, heptachlor, lindane, mirex, kepone, 2,4,5-T, 2,4-D
- chlorinated compounds such as dioxins
- polynuclear aromatic hydrocarbons
- heavy metals – arsenic, cadmium, lead, mercury
- bacteria, viruses, protozoa, parasitic worms, fungi
- miscellaneous – asbestos, petroleum products
- industrial solvents.

And that's just the sludge!

Once the effluent has been extracted from the sludge, it is then available for processing into a clear solution. It is the chemical composition of this innocuous clear liquid that concerns many scientists. Recent research reveals that pharmaceutical drugs and chemicals are turning up in water supplies all over the world – in surface water, groundwater, and drinking water at the tap. These chemicals include drugs associated with medicine, agriculture and the pastoral industry: hormones, antibiotics, painkillers, tranquillisers and toxic chemotherapy mixtures.

A pioneer study in the 1980s, by Mervyn Richardson of Thames Water in the UK, found that the River Lea in north-east London was carrying detectable concentrations of 170 different drugs, ranging from aspirin to morphine derivatives. He estimated the annual load of these drugs in the river to be more than 1 tonne, implying concentrations of up to 1 part per billion.

Where had the cocktail come from? Treated effluent from upstream sewage-treatment plants seemed to be the logical answer.

Other more recent scientific findings confirm Richardson's work. They have also found many new toxic chemical substances in treated sewage effluent, including the group known as nonylphenols.

Nonylphenols are used in the manufacture of common laundry detergents, household cleaners, pesticides and plastic softeners. Because nonylphenols interfere with endocrine function, they are known as endocrine-disrupter chemicals (EDC), clinically defined as "exogenous substances that change endocrine function and cause adverse effects at the level of the organism, its progeny, and/or (sub) populations".[5]

Chemicals that mimic the effects of the female hormone oestrogen have been linked to sexual abnormalities and disorders in animals and humans. Manifestations include panthers with undescended testes, alligators with abnormally small penises, bisexual seagulls, hermaphrodite fish, and laboratory rats with scrambled reproductive systems. The effects of these rogue hormones on humans include, as well as the above, a lowered sperm count and consequent difficulty with conception, a doubling of testicular and prostate cancer in men and an increase in breast cancer in women.

Both chlorine and organochlorines produce synthetic oestrogens, and some scientists are investigating the effects on humans of chlorine compounds widely used in agriculture, industry, and water and wastewater treatment.

Scientific research to date indicates that endocrine-disrupting chemicals can induce biological effects in humans at concentrations as low as parts per trillion. They cannot be removed by conventional water treatment processes.

Studies undertaken in Australia in 2003 and 2004 identified a wide range of pharmaceutical compounds including analgesic, anti-inflammatory, anti-convulsant, anti-hypertensive, anti-opiate, lipid-regulating – and even illicit drugs, in municipal sewerage. "Potential effects of ingestion or agricultural application of these complex low-concentration mixtures remains largely unknown. There exists a huge array of biological compounds many more of which have not been identified but are likely to be present".[6]

Many of the toxic or biologically active EDCs in sewage effluent were undetectable and/or unknown 10 years ago. A recent example is the potent carcinogen, nitrosodimethylamine, known as NDMA. NDMA has been found in chlorinated effluent intended for reuse and by implication existed in water recycled previous to its discovery. NDMA is believed to be formed during the disinfection process.

Since the increased rate of detection of these compounds is the result of advances in water analysis, it is likely that many more such substances will continue to be identified.

Bearing in mind the possibility of the long-term health, genetic and fertility effects of some of the toxins found in effluent, it is virtually impossible to establish its safety. The Singapore NEWater project, for example, had a two-year "Health Effects Testing Programme" to "address the potential health impact of unidentified contaminants."

They tested the effluent on fish and mice – for two years. And on that basis they pronounced it "safe"? Cancer, genetic and fertility defects take many years – or even generations – to manifest.

It is the population of Singapore who are the *real* subjects of this experiment, not the mice and fish. No wonder the sales of bottled water in the city skyrocketed after the introduction of NEWater.

Highly infectious *Cryptosporidium parvum* oocysts were recently reported in 40% of the final disinfected effluents from six water-recycling facilities in the USA. Of the microorganisms, enteric viruses present the greatest potential health threat in the reuse of effluent. These are excreted in high numbers by infected individuals and are difficult and expensive to remove, so increased rates of water recycling will always involve an increased risk of waterborne disease.

We still have a long way to go before we can confidently assert that recycled water is, firstly, safe to drink or secondly, safe to release into the aquatic environment as a substitute for natural environmental flows.

This is another use being proposed for recycled water. Since it is now accepted that rivers and streams need a certain minimum volume of water to ensure their "health", some "experts" are suggesting that treated effluent could suffice. Is this the sort of future we want to offer to our children: rivers of effluent?

It's already happening.

Up to 93% of the annual flow of Victoria's Thomson River is diverted into Melbourne's storages. Downstream from the Thomson Dam, the river becomes turbid and laden with excess nutrients, and often ceases to flow at all in its lower reaches during summer. The following editorial from Melbourne's *The Age* newspaper (28 April, 2004) gives a helpful perspective on the issue of substitute environmental flows.

> A $1 billion-plus proposal floated this week by Deputy Premier John Thwaites to pump water from Melbourne's treated sewage back into the Thomson River is an indication at least that the State Government is taking the future of the state's water supply seriously. That may be a moot point for the people of Gippsland. Having relinquished their water to the city, whether they would find it acceptable to have it returned to them having passed through the digestive systems of the people of Melbourne is certainly debatable. Resentment at the proposal is surely understandable. Mr Thwaites has acknowledged that any plan to pump treated water into Melbourne's supply would be "politically unacceptable". But is it any more acceptable to pump it back to Gippsland? There must also be environmental concerns about passing treated water back into a natural river system, rather than diverting it more directly into obvious agricultural or industrial uses.

Interesting, isn't it? Adding effluent to Melbourne's water supply is "politically unacceptable" but it's okay for the drought-stricken farmers of Gippsland. It's not dissimilar to a controversial proposal that created a furore in the US in 2002.

When the Californian city of San Diego tried to establish an 80 mega-litres per day indirect potable reuse facility, a vocal group of the city's population rejected the project on health grounds. Their opposition took on racial overtones when it was pointed out that treated waste water from wealthy neighbourhoods would be distributed to poorer urban communities who would thus be compelled, as they put it, to drink 'the effluent of the affluent'. Critics of the project brought large display boards to public meetings that showed directly piped connections between the toilet and the kitchen tap. The project was shelved.

A similar proposal in southern Queensland met with the same fate.

In "Demonstration, the Solution to Successful Community Accept-ance of Water Recycling", Howard Gibson and Nick Apostolidis note that "a medium-sized local government was defeated at election for attempting to promote indirect potable reuse". The article continues:

> Along Australia's Sunshine Coast, consideration by the Maroochy and Caloundra councils to augment dwindling domestic water supplies with recycled water was modified after strong opposition from the local community. Community action groups used media releases to warn the community about the potential shrinking and deforming of male sexual organs because of gender-bending hormones in wastewater. An internet website was also established by one resident's action group, the Rivermouth Action Group, to spread news of the reuse proposal scheme citing the feared ill effects of using recycled water in the US, and how the proposal would be the first of its kind, reusing infectious hospital, abattoir and industrial waste for potable purposes.[7]

In the final analysis there are several obstacles to be overcome if recycling is to provide a substantial alternative to conventional water supplies.

The safety of recycled water – even for environment flows, or for irrigation purposes in catchment areas – has yet to be conclusively established. Indeed, there is considerable evidence that there are potential health risks to consumers and to the aquatic environment.

Are these risks an informed community is prepared to take?

Ultimately, I suppose it depends on how thirsty we get in the years to come.

Silent Springs
and Depleted Aquifers

Now the stock have started dying, for the Lord has sent a drought;
But we're sick of prayers and Providence – we're going to do without;
With the derricks up above us and the solid earth below,
We are waiting at the lever for the word to let her go.
Sinking down, deeper down,
Oh, we'll sink it deeper down:
As the drill is plugging downward at a thousand feet of level,
If the Lord won't send us water, oh, we'll get it from the devil;
Yes, we'll get it from the devil deeper down.

> – A.B. (Banjo) Patterson, "Song of the Artesian Water" 1895

As groundwater tables in many parts of the world are declining, great rivers are utilized to the point where they no longer reach the ocean, more soils are taken out of production due to erosion, waterlogging and salinisation than virgin soils are brought under the plough, and clean water, free from industrial pollution and disease-carrying organisms, is a scarce resource, we have come to a point where water scarcity is increasingly perceived as an imminent threat, sometimes even the ultimate limit to development, prosperity, health and even national security.

How close are we to those limits?

> – Leif Ohlsson, *Hydropolitics: Conflicts Over Water*
> *as a Development Constraint*, 1995

GROUNDWATER IS ANOTHER RESOURCE to which cities and towns turn when surface water supplies begin to dry up. Between 1985 and 1997, the extraction of groundwater in Australia increased by a massive 90%, and during the past eight years, that use has continued its steady upward trend. In 1990, the Australian Water Resources Council estimated that groundwater was the main source of drinking water for a million people. The *Australia State of the Environment* report in 2001 tells us that four million people depend on diminishing and over-exploited groundwater for their domestic supplies. This growth in use reflects current international trends.

Groundwater is formed when rain or surface water percolates through the soil into underground reservoirs called aquifers. Aquifers are geological formations of porous materials such as sand and gravel, or spaces between subterranean rocks. It may come as a surprise to learn that most of our usable water is not visible on the surface: around 97% of the planet's fresh water supply is stored in underground aquifers.

Much of it is thousands or even millions of years old.

Some aquifers can refill quite rapidly if the ground surface is porous, while others may take millennia to recharge. This is an important distinction because deep old fossil aquifers, once depleted, may not be replenished until the year 3000, and that's a long way off. The average residence time for groundwater is 1400 years, as opposed to 16 days for river water.

Drought is also a factor. Low rainfall means low rates of recharge. During the past 30 years, Western Australia's rainfall decreased by between 10 and 20%, which in turn diminished groundwater replenishment by 40–50%.

It's important to always keep in mind that groundwater, rainfall and surface water are all connected; all are part of the same cycle. The quality and quantity of one aspect is an indication of the overall condition of the entire resource.

The actual extent and size of large aquifers and the volume of water they contain are difficult to estimate accurately. Hydrologists can only make educated guesses. "Sustainable yield" is the term used when trying to estimate how much water can be taken from an aquifer so that the rate of extraction equals the rate of recharge.

Once the demand for water exceeds the sustainable yield, the gap between extraction and recharge grows exponentially every year. During the first year, the water level will fall very little, but each year thereafter the annual drop in the water table will be larger than the year before.

And those levels are falling rapidly around the world: as the global population multiplies, so does its dependence on groundwater. As extractions increase, wells and aquifers are beginning to dry up – in the USA, China, India, Australia and the Middle East.

The Sri Lanka-based International Water Management Institute points out that this has dramatic implications for future production and industrial growth as well as water supplies.

> Many of the most populous countries of the world – China, India, Pakistan, Mexico and nearly all the countries of the Middle East and North Africa – have been literally having a free ride for the past two or three decades by depleting their groundwater resources. The penalty of mismanagement of this valuable resource is now coming due and it is no exaggeration to say that the results could be catastrophic for these countries, and given their importance, for the world as a whole.[1]

In western Yemen, for example, where the rate of extraction in the Sana Aquifer exceeds the rate of recharge by a factor of five, the water table is falling by six metres every year. World Bank projections indicate that this water supply for the nation's capital, and its two million people, will dry up by 2010. Test wells drilled two kilometres deep have failed to find water.

Other countries are facing similar shortfalls. Wells in Gujarat in India have seen water tables gradually drop from 15 metres to 400 metres since 1975. In the US, underground supplies available for irrigation are shrinking at the rate of 2% every year, or at least 20% every decade.

The Great Artesian Basin is Australia's largest groundwater resource, lying beneath one-fifth of the continent. Its estimated capacity is 8700 million gigalitres of water that has been there for a million years or more, but vast quantities have been allowed to go to waste. Some bores in the Basin have been flowing constantly since the 1870s, the water fed into long open drains for stock to drink.

Most of it evaporated.

In 2003, there were still 892 uncontrolled bores flowing freely 24 hours a day. This water wastage has been costed at $14 billion in lost future national production each year. However, the Basin is so remote from our coastal centres of population that its waters remain far beyond their reach.

So our focus narrows to the groundwater that may help to alleviate the current urban water crisis. Here the news is not so good either. The limiting factor is pollution.

Unlike surface water, groundwater is not exposed to the natural purifying effects of air and sunlight that evaporate volatile compounds and help to disperse organic contaminants. Aquifers have few microorganisms capable of breaking down organics and neutralising their effects. Consequently, groundwater is highly susceptible to contamination from all of the activities that take place above it. Agriculture, mining, industry, toxic waste dumps, landfills, leaking sewer pipes, septic systems and underground storage tanks are the main polluters.

Think of our urban landscape as the top layer of the soil profile, and then consider that everything we add to that soil, or spray on our plants, or bury in the ground will have an impact on the water below. Groundwater under urban areas is frequently contaminated with the residues of fertilisers, herbicides, pesticides and insecticides, among other things.

Australians spend around one billion dollars each year on these highly concentrated and persistent poisons, which we then apply liberally within our cities, towns and water catchments. An estimated 20–40% of this toxic load is used in urban areas, says Dr Kate Short in *Quick Poison, Slow Poison: Pesticide Risk in the Lucky Country*.

Four per cent of this is used by the pest control industry and the remainder is divided between private and public users for non-agricultural purposes, particularly spraying for weeds.

Major herbicide users include local councils who spray roadside verges, ovals, commons, parks, waterways and wetlands to control weeds and insects like sandflies and mosquitoes. Telecom and the railways use herbicides to control weed growth on lines and embankments, under power lines and around sub-stations, and the various state forestry commissions aerially spray vast tracts of clear-felled native forests to prepare the land for mass plantations. Other big users are the statutory authorities charged with noxious weed eradication. Every year they spray tens of thousands of litres of herbicides onto pasture, range and forestry lands, often with little regard for public safety.[2]

And when it rains, residues of these weapons for the mass destruction of nature slowly begin to percolate down to the groundwater that in turn becomes the rivers and streams that we rely on for our drinking water.

In ecosystem terminology, groundwater is sometimes referred to as a sink, by virtue of its position, and gravity being what it is. A natural sink is a medium for the decomposition, detoxification and the accumulation of wastes, human, animal and mineral.

The most dangerous of these is industrial waste. In terms of volume and toxicity, mining and industrial wastes constitute the biggest single threat to surface and groundwater supplies worldwide.

There are endless examples.

The intricacies and legal complications of groundwater contamination are detailed in the movie *Erin Brokovich*, the story of a true incident involving the release of chromium-doped cooling water to unlined ponds from a large power plant in Hinckley, California. This tainted the groundwater supply of the local residents and resulted in a large lawsuit against the company. Since most companies manage to avoid detection and/or prosecution, the case attracted considerable media interest.

Australia's most serious groundwater contamination incident is still unfolding beneath the industrial suburbs of south-eastern Sydney.

A massive toxic plume, created by leakage and "releases" of carcinogenic chlorinated hydrocarbons by the chemical giant Orica, has contaminated the Botany aquifer. Orica has admitted that the process continued for thirty years.

In 1990 the NSW Government was made aware of the contamination, but apparently took no action to warn bore users until October 2003, when local residents were told at a public meeting that their backyard bores were contaminated with levels of ethylene dichloride up to 1.6 million times higher than the safety level in drinking water standards.

A representative of the Environment Protection Authority (EPA) told residents that the chemical had been "released" by Orica between the 1940s and the mid 1990s: "There were leaks and discharges and spills [by Orica into the groundwater] from what we would regard today as poor environmental management," he said.

What a massive understatement.

A city's water supply is poisoned by years of criminal neglect and illegal discharge. What sort of punishment should such a company receive? What do you think would compensate for this loss? What would be appropriate?

The fine was $320, less than the cost of a speeding offence. Hard to believe, isn't it?

The spokesperson continued, "Once it [the chemical] is in the groundwater, it is difficult to deal with."

The Daily Telegraph's "Urban Jungle" reporter Mark Skelsey described the mood of that 2003 meeting as "ugly". He continued: "Residents who have sunk shallow bores say they have received no information over the years about this plume."

Eighteen months later, after several failed attempts at remediation, the two-square-kilometre toxic mass was still moving slowly and inexorably towards Botany Bay, "an area of incomparable environmental, cultural and historical significance".

University of NSW hydrogeologist Jerzy Jankowski, said the toxic plume was the biggest in the southern hemisphere, and it would be a tragedy if it reached the bay. Dr Jankowski said the bay's edge would become a "contaminated site" where no human contact with the water would be permitted. "It would be a complete disaster for marine life, such as oysters, seaweed and anything which lives in the sand," he said.

The Orica site is just one example of the damage that can be caused by industrial waste – one of an estimated 80,000 contaminated sites that lie underneath urban Australia, beneath our cities. How many there are in rural areas is not known.

The Victorian EPA keeps a comprehensive register of 10,000 contaminated sites in that state. Sites on the EPA's Priority Sites Register contain dangerous levels of arsenic, cyanide, asbestos, lead, petrol and other chemicals "where pollution of land and/or groundwater presents an unacceptable risk to human health or to the environment". Because these contaminants may take years or even decades to percolate down, their effects may be delayed, so that people may end up drinking groundwater containing contaminants that have long since been forgotten about.

The NSW EPA also notes that almost all of the older petrol stations have leaking tanks which are polluting underground water supplies.

It doesn't take much fuel to poison groundwater. Petrol contains benzene, toluene, ethyl benzene and xylenes. These compounds are both soluble and mobile in water and can cause cancer at very low concentrations. The maximum permissible level for benzene in drinking water is 10 parts per billion. A 1991 CSIRO study estimated that one-fifth of Perth's 500 service stations had leaking underground tanks polluting the city's groundwater supply. This proportion would probably be similar in other areas and cities.

Mining is another activity that affects groundwater. Arsenic from old gold-mines, lead, chromium and other metals are not biodegradable and can linger indefinitely in groundwater.

Acidity is another emerging groundwater issue.

At normal levels of acidity, soils can absorb pollutants. Soils bind with toxic metals in particular and retain them so that they don't poison surface and groundwater and, ultimately, human beings. But when conditions become more acidic these bonds are dissolved, as W. M. Stigliani points out in *Chemical Time Bombs*:

> As soils acidify, toxic heavy metals, accumulated and stored over long time periods ... may be mobilized and leached rapidly into ground and surface waters or be taken up by plants. The ongoing acidification of Europe's soils from acid deposition is clearly a source of real concern with respect to heavy metal leaching.[3]

The *Australia State of the Environment 2001* reports that the situation in this country is also deteriorating. It identified:

> ... increasing trends in water acidity and the area of land affected by soil acidity ... since 1991 the area of land affected by acid soils has increased by 13 million hectares to 47 million hectares... High water acidity may lead to increased availability and movement of pollutants as well as changes to the chemistry of rivers and streams.[4]

What does this mean to us in the Twenty Thirst Century? To what extent will all of this additional contamination further decrease our available unpolluted water supply?

The widespread acidification of soil and water is the consequence of the overuse of fertiliser, fallout from industrial emissions and/or the depletion of groundwater. Our entire environment is progressively becoming more acidic. As soils acidify so does the ground and surface water flowing through them, and this in turn has serious implications for human health.

Let me give you a local example.

The Somersby Plateau is a small farming community on the Central Coast of NSW which depends on groundwater for its supplies. In 1999 water could easily be obtained from 30–60 metres below the surface. Now in 2005 it is necessary to drill down 150 metres to locate water. As the volume of water in the aquifer diminished, the acidity of the water began to increase, dissolving high levels of naturally occurring aluminium from the surrounding rocks and soils.

A Central Coast Health Study revealed that some bores on the plateau had both high acidity and high aluminium levels. Some of Australia's most popular brands of spring water come from bores at Peats Ridge. Laboratory testing of a selection of commercial spring waters revealed that Peats Ridge Springs was the most acidic (lowest pH) and had the highest aluminium content. The story went to air on Channel 7's *Today Tonight* on 1 November, 2004.

I thought there was some irony in the fact that, while consumers in nearby Sydney were buying Coca-Cola Amatil's *Peats Ridge Springs* brand because they were concerned about the quality of their tap water, they were unknowingly consuming spring water containing levels of aluminium 200 times that of Sydney Water's finest beverage. The only obvious health benefit to consumers was the exercise involved in lugging the heavy spring water containers home!

Around 500 million litres of bottled water – a cost of $525 million – is consumed by Australians every year, much of it groundwater.

The bottled water industry is unregulated and subject to little government control. In some areas as the drought intensifies, competition is growing between local residents and powerful bottled water companies with a documented history of pumping aquifers dry, pocketing the profits and moving on.

> In most countries, water corporations source underground water for their bottled water industry without prior determination of the right to harvest from such sources and an assessment of the adverse impacts that such harvesting will have on the environment, farms and households that are also reliant on the same source for water. They seek formalisation of water rights that will enable them to acquire monopoly rights. The formalisation they promote is one that seeks to exclude communal ownership and penalise indigenous and farming communities.
>
> – "Water: Public Supply or Private Profit?" *Choice*, May 2004

This practice is attracting criticism, both in Australia and overseas.

In a landmark decision in December 2003, the Kerala High Court ruled that Coca-Cola must close down pumping operations at its massive 16-hectare bottling plant in southern India because it was threatening the security of surrounding groundwater supplies. This was the culmination of two years of protests by farmers who complained that their wells were drying up. Coca-Cola denied the charge, claiming it was untrue.

The court found that groundwater was a national resource that belonged to the entire society. "Groundwater under the land of the company does not belong to it," said Justice K Balakrishnan Nair.

> Every landowner can draw a 'reasonable' amount of groundwater
> which is necessary for its domestic and agricultural requirements.
> But here, 510,000 litres [110,000 gallons] of water is extracted per
> day, converted to products and transported, thus breaking the
> natural water cycle.
> – *The Guardian*, 19 December, 2003

The judge pointed out that underground water was public property and that the state had a duty to act as trustee for its protection. The government had no power to allow a private party to extract such a huge quantity of groundwater, which could result in its drying up, he concluded.

Will Australian courts follow India's lead? The issue may well surface here in the near future.

"Fears grow over tapping water table" was the heading of an article in *The Gold Coast* Bulletin 28 August 2004. Journalist Ryan Ellem reported that:

> Tamborine Mountain residents are concerned that water carriers
> boring into local supplies for multi-national soft-drink makers will
> bleed the community dry. The underground water body has not been
> declared an aquifer by Natural Resources Minister Stephen Robertson,
> allowing anyone to bore into the supply and sell it. Water taken
> from the body is not monitored, leaving residents in the mountain's

unreticulated villages anxious about the availability of water, and price hikes if they have to source carriers from other regions. Most residents in the area use water carriers to fill up their household tanks when they're getting low but this was now becoming difficult due to increased demand. One domestic water carrier had 700 people on their books and a long waiting list.

Local councillor Vanessa Bull told me she believed that residents should have first access to the water. Councillor Bull also confirmed an earlier report which suggested that at least 150,000 litres was leaving the mountain every day where it was carted by tankers to Coca-Cola's Brisbane plant to be sold under the Mount Franklin label. Court action was a possible option as a last resort, Councillor Bull said.

Alec Wagstaff, Corporate Affairs Manager for the bottler, Coca-Cola Amatil, denied that over-pumping was a problem. "We conduct independent hydrological studies of all the water sources we use for quality and sustainability purposes. We carry the same concern as the community in terms of supply, because we want to stay around producing for a long time," he said.

Mr Wagstaff's concern about the sustainability of the resource is difficult to reconcile with the company's earlier mission statements. In its 1993 annual report, Coca-Cola outlined a grand vision for the future:

> All of us in the Coca-Cola family wake up each morning knowing that every single one of the world's 5.6 billion people will get thirsty that day. If we make it impossible for these 5.6 billion people to escape Coca-Cola, then we assure our future success for many years to come. Doing anything else is not an option.

Coca-Cola's thirst is insatiable. Its global water empire includes more than 20 brands in more than 100 markets. Their Australian brands include Pump, Neverfail, Mount Franklin and Peats Ridge Springs.

In April 2005, Coca-Cola Amatil requested – and was granted – an increase in the Peat's Ridge Springs water allocation, from 25 million litres to 66 million litres annually, by the NSW Department of Planning and Infrastructure. The development application was refused by Gosford Council on the grounds that it would reduce the base flow in the Central Coast drinking water supply by 41 million litres each year.

The base flow is the proportion of groundwater seeping into streams and rivers. In many areas, this base flow contributes around 50% of the average annual stream flow. So as unsustainable levels of groundwater extraction reduce the volume of base flow, they will also reduce the amount of water flowing in nearby rivers, streams and dams.

Water off a Coke chairman's back

The national interest argument has been used on oilfields and uranium mining but I've never heard it in reference to bottled water.

The link was made somehow by David Gonski in a speech he gave on corporate responsibility to the Institute of Chartered Accountants in Sydney on Friday.

The chairman of Coca Cola Amatil seems to think it's not easy to get a drink of water in this country.

The subject came up because Coke is meeting some opposition to expanding its water bottling plant near Gosford on the NSW Central Coast. 'My view is that it's quite a good thing for people to drink water and I think we should make it as easy for them to do it as possible,' Gonski said. He said people would not have access to water without his company.

The serial director then stated the obvious.

'We are taking water from our Peats Ridge spring and putting it in bottles so that people can drink it. 'The opposite of that is we leave the water there so they can't', said the man who usually concerns himself with higher thoughts than these.

Coke wants to increase output at its Peats Ridge bottling plant from 25 megalitres to 66 megalitres a year.

However Gosford City Council knocked back the proposal at a meeting last week, citing concerns that it would further reduce the city's dwindling water supply which could actually make it tough to get a drink of water.

– *Australian Financial Review* 18 April, 2005.

The Central Coast's dam water levels were, at that point, down to 24% of capacity and falling – yet the bottler's allocation was increased by government planners.

Announcing that Coca-Cola Amatil would fight the council's decision in the Land and Environment Court, Alec Wagstaff said the council's response was more emotional than scientific and reiterated Coca-Cola's heartfelt commitment "to managing the resource sustainably" (*Central Coast Express Advocate*, 4 May, 2005).

What sort of planning prioritises corporate profits at the expense of public drinking water in a time of drought when residents are experiencing water restrictions?

As Mr Wagstaff observed, it's quite obvious that Coca-Cola will be around for a long time. Whether the shrinking water supplies of the Central Coast or Tamborine Mountain will last the distance is another matter.

Unsustainable groundwater depletion, whether it is sanctioned by government or not, amounts to theft of the communal supply. It's a crime that should attract the same penalty as any other stealing offence.

While many fine political speeches are made about the necessity for groundwater conservation and management, out there in the real world – out-of-sight, out-of-mind, deep under the earth – unnecessary and excessive consumption is sucking the lifeblood out of ancient storages, some of which will never be replenished during this century.

Effective groundwater regulation is a national issue, which demands an urgent and considered response.

> Governments and communal authorities need to institute or strengthen groundwater regulation. A classic common-pool resource, groundwater is susceptible to overuse because the collective impact of each user acting out of self-interest is the depletion of the supply of all. Sustainable use of renewable aquifers requires that total withdrawals not exceed the level of replenishment. As researchers at the Sri Lanka based International Water Management Institute point out, however, "Nowhere in the world do we find such an ideal regime

actually in operation ... Precious little is being done to reduce demand for groundwater or to economise on its use."[5]

When will our governments, both state and federal, begin to take responsibility for protecting vulnerable community supplies from over-exploitation?

Never, it seems, if it interferes with jobs, economic growth or company profits. In the end it's going to be a matter for us as a society to decide whether we agree with this principle, if you could call it that.

This country cannot sustain the current levels of groundwater use, let alone the growth projected for the next two decades. If the Australian population and its demands increase, and unpolluted surface water supplies diminish at the current rate of depletion, some areas dependent on groundwater will face the prospect of running out of water altogether. Where water levels are already dropping, digging deeper wells will tie up more money and resources; pumping water from these lower depths will increase energy use and, consequently, costs. And water from lower aquifer levels is often of poorer quality because naturally occurring contaminants such as aluminium, arsenic, fluoride and radon are more common at greater depths, as the Earth's higher internal temperature at these levels dissolves more of these elements into solution. Changes in pH due to depletion also result in the leaching of toxic substances into groundwater.

Unsustainable depletion of coastal aquifers can disturb the equilibrium and allow the intrusion of salt water, making the contents of the underground supply unsuitable for human consumption.

Over-pumping of groundwater has caused land disturbances, where the land above the empty aquifer may crack open or even subside, sometimes quite dramatically.

And finally, particularly in times of drought when groundwater may constitute a high proportion of surface water flows, depletion may result in drying or diminishing the potential water levels of wetlands, rivers, streams, lakes and dams. In some areas in the world, over-pumping has already caused rivers to dry up, ground to collapse, cracking house foundations, and trees and groundcover to die.

Economists describe these calamities as "externalised costs" or "externalities", because they are not an expense for the pumping company.

In other words, the benefits of over-pumping groundwater flow to the individual pumper, but the costs of environmental degradation are shared by the community who are dependent, either directly or indirectly, on the aquifer.

Pollution and unsustainable resource depletion, whether its by farmers, irrigators, water authorities or water bottlers, is an issue that ultimately affects us all.

In the Twenty Thirst Century it may be the crucial factor that determines our survival.

No Water Dreaming
The Sydney nightmare

Sadly we believe the world will experience overshoot and collapse in global resource use and emissions ... The growth phase will be welcomed and celebrated, even long after it has moved into unsustainable territory (this we know because it has already happened). The collapse will arrive very suddenly, much to everyone's surprise. And once it has lasted for some years, it will become increasingly obvious that the situation before the collapse was totally unsustainable. After more years of decline, few will believe it will ever end. Few will believe that there once more will be abundant energy and wild fish.

Hopefully they will be proved wrong.

> – Dennis Meadows and Jorgen Randers,
> *Limits to Growth: The 30 Year Update*, 2005

I wake up in the morning thinking there are lots of times when people have woken up feeling like this, like Old Testament prophets ... There will be conditions not seen in 40 million years ... I try to find a way out of it but I can't. It's life changing to realise what is going on.

> – Dr Tim Flannery, speech to the
> Sydney Futures Forum, 18 May, 2004

EARLY IN THE MORNING of Wednesday 21 July, 2004, I woke up bathed in sweat and overcome by a terrible fear. I looked at the clock. It was 4 am. Outside my window the familiar sounds of a sleeping metropolis. I went to the bathroom and turned on the tap just to reassure myself that the nightmare I had left behind was not a reality.

I dreamed that Sydney had run out of water.

It was a very disturbing experience: A sudden unexplained power surge had brought down the energy grid some days previously, leaving the city without power and in chaos. No lights, no cooking stove, no refrigerator, no lifts, no radio, no TV, and now, no water.

My partner and I had queued for hours to buy a two-litre ration of water in the dark recesses of a Coles supermarket, a queue made longer by the fact that the electronic tills and scanners were not working. It was very hot and airless without air-conditioning. Everyone was sweating. The smell of fear was in the air. It didn't smell good.

Back in the apartment the situation was rapidly becoming untenable. The dirty dishes in the sink were not an issue, but the toilet was rapidly filling with waste and it wasn't going anywhere.

The smell in the bathroom was overwhelming. It was time to go.

We packed some fruit and a few belongings in backpacks and started walking north.

The roads were full of abandoned cars, trucks and Toorak tanks. No public transport meant that everyone with a vehicle had decided to take off for greener pastures. But the electronic petrol pumps were not working. Some enterprising service station proprietors had con-

---

The North American blackout of 2003 affected 50 million people in Canada and the US.

When pumping stations failed, drinking water was distributed by the National Guard.

Advice to consumers – Buy bottled water and put it away. No matter how much you think you'll need, you'll probably need more, so go for multi-gallon jugs. After the power fails but before water tanks run dry, it's also a good idea to fill your tubs. (*Time*, 25 August, 2003)

nected hand-driven ratchets, but the process was slow and queues stretched for miles.

In the absence of traffic lights, gridlock built upon gridlock. Cars overheated and stalled. Others broke down. Eventually, traffic ground to a halt. After spending an hour or two in the hot sun, many decided to abandon their cars and continue on foot. This meant there was no hope that the traffic situation could improve in the immediate future.

We joined the thousands of people with dogs and children, pushing shopping trolleys, prams and strollers piled with food and possessions, all walking, heading north. By mid-morning we were still in the northern suburbs. The sun was hot and the last of our water was gone.

We were very thirsty …

But of course it was only a dream, I told myself. Such things could never really happen – not in Sydney. Nevertheless I could not get back to sleep. What if the dams really did dry up or the grid went down, and the city was left without water?

The Great Drought of 1934 which lasted for eight years had come close to shutting down Sydney's water supply. That was why Warragamba Dam was built – to make sure we would never run out.

And now Warragamba was down to 45 per cent of its capacity and falling. Was it half full or half empty? Were we three years into an eight-year drought, or seven years into a 10-year one? Doubts preyed on my troubled mind.

At 9 am I rang Colin Judge at Sydney Water to ask him if there was a contingency plan if, for whatever reason, Sydney should happen to run out of water.

Colin laughed. "You've got me there John," he said. "I think you'll have to ask Frank Sartor."

Yes of course. The busy ex-mayor of the City of Sydney who was now the Minister for Many Things, including Energy and Utilities,

Science and Medical Research, as well as being (according to his letterhead) Minister Assisting the Minister for Health (Cancer) and Minister Assisting the Premier on the Arts. Sydney's water supply was just another of his many, many responsibilities. One hoped it was high on his long list of priorities.

Perhaps Frank could reassure me.

Frank was busy when I rang. Well you would be, wouldn't you, with all those portfolios to worry about, but Michelle – one of his impatient young assistants told me she was sure there was a plan, but when she asked around the office, no one seemed to know what it was.

"Can I take your number and get back to you?" said Michelle with a tone that implied that she had more important things to think about.

I never heard from Michelle again.

I called Sydney Catchment Authority, responsible for the Warragamba Dam and its catchments. Their spokesperson was sure there was a plan, but she didn't respond positively to my request for a copy.

"Perhaps you should talk to Bob Debus," she suggested.

Ah, yes – Bob Debus, the NSW Attorney General *and* also Minister for the Environment. Aren't we lucky to have such multi-skilled people looking after all these things for us?

As Minister for the Environment, Bob was responsible for the Catchment Management Authority who was responsible for the dam.

Perhaps Bob could set my mind at rest.

But, alas, Bob was also too busy to talk and so was Chris Ward, his assistant. Someone at the office suggested that Craig Knowles might be able to help.

Now I'd had a few meetings with Craig during the Sydney water crisis in 1998. Back then he was looking after Sydney Water, but now he'd moved on to become Minister for Infrastructure and Planning *and* Minister for Natural Resources.

Kathy, his assistant, was very busy. In response to my enquiry, she faxed me a press release dated 8 June, 2004.

It didn't say anything about a contingency plan.

What it said in part was: "Irrespective of whether we live in the city or in the bush, we all have to work smarter to save this continent's most precious natural resource – water."

Slowly I began to realise that there was, in fact, no plan.

Here we were, sitting ducks.

Australia's largest city – its glittering commercial jewel, the power-house of the economy and social hub of the nation – and no obvious evidence so far of any effective strategy to deal with the possibility of the city running out of water. And that possibility looming large on the horizon.

I called around the multi-skilled circle of government ministers once more and challenged them and their staff to either produce a plan, or admit that none existed. I ranted and raved and made a nuisance of myself. Then I invoked the Freedom of Information Act and sud-denly, as they say, the dam burst.

A plain brown envelope arrived, courtesy of Sydney Water. Here it was. The plan. At last.

I tore it open.

The Drought Response Management Plan 2002–2012 (revised January 2003) has a thought-provoking quote on the cover: "We never know the worth of water till the well is dry."

That didn't augur well for the present situation.

The management plan runs to 54 pages, with 30 pages of appendices. It begins as follows: "A prolonged period without precipitation is of great concern to Sydney because it relies on surface water for bulk water supplies."

Yes indeed. Can't argue with that, can we? The document continues:

> Consequently, drought response measures can be devised and implementation procedures defined to ensure an uninterrupted supply of water sufficient to supply customer's essential needs and at the same time minimise adverse impacts on the quality of life, environment and the economy.
>
> The drought response measures form the core of Sydney Water's Drought Response Management Plan (the Plan).

So this is it. The core. This is "The Plan".

There's some talk about the models used to predict future demand.

We learn that there are two models to choose from. One produced by the CSIRO in 1998 focuses on climate change, and forecasts – quite accurately as it turns out – that "the effects of climate change are likely to lead to significantly higher rainfall in coastal areas compared to inland NSW. **The Warragamba Dam catchment is likely to experience drier conditions than the Sydney Metropolitan Area** …" [emphasis mine].

However, Sydney Water in its wisdom decided to choose a more optimistic model, and as a result the CSIRO climate change forecast "was not factored into predictions of future water demand".

Why not, one wonders?

The Plan was not looking good so far.

I hurried over pages of waffle about restrictions and exemptions for business, industry, commerce and government, until finally, I found what I was looking for on page 47, under the heading "Contingency Plans".

Acknowledging that Sydney has never before experienced such a severe water shortage, The Plan continued: "We must be prepared for the worst situation, however, and have available some alternatives to ensure there is enough water for the most basic needs of Sydney."

There are five alternatives that the expert consultants and water boffins put forward to save the city from dying of thirst.

The first I christened Frank's Ark, because of Frank Sartor's enthusiasm for desalination. The idea was to have a floating desalination plant "on a ship moored off shore and the fresh water is piped to the land". Now this might work for a village with a couple of hundred people, but to make a significant contribution to Sydney's water supply, such a ship (which has not yet been built) would have to be bigger than Noah's Ark. It would require a vast amount of power to operate, and massive pipelines leading from it to the shore. I can just see all the Double Bay housewives queuing up at the outlet with copper buckets balanced on their heads.

I don't think a serious person could really call that a viable solution. So I can move on to the next emergency option: Groundwater.

Tapping into the Botany aquifer appears as the first proposition. Now I don't know if anyone at Sydney Water reads the local papers such as *The Sydney Morning Herald* or *The Telegraph*, but if they had they might have noticed that the Botany aquifer is the site of a multi-million dollar clean-up because it is heavily contaminated with cancer-inducing chlorinated hydrocarbons from an old industrial site. Any large-scale pumping operation would impact on the possibility of remediation and cause the toxic material to spread even further.

So lets forget about the Botany aquifer for the moment.

Unfortunately, the other aquifers underneath Sydney are likely to be equally unsuitable for drinking purposes as a result of two centuries of pollution slowly filtering down from above – toxic sites, waste dumps, deliberate pumping of industrial waste into aquifers to avoid costly disposal, rubbish tips, and thousands of petrol service stations with their leaking underground tanks.

According to CSIRO estimates, at least one-fifth of all service stations in Sydney have leaking underground fuel tanks. There may be more: no one knows for sure. Petrol contains benzene, trichlorethelene and perchlorethelene, all of which are known to cause cancer even at very low concentrations. The World Health Organisation has set a maximum permissible limit of 10 parts per billion for benzene in drinking water.

Other groundwater options include pumping out aquifers on the NSW North and South Coast and trucking the water back to Sydney.

And then what? More queues with buckets?

And by the way, residents of the North and South Coast have their own water shortages to deal with and might not take kindly to the idea of having their community supplies depleted in order to make up for Sydney's shortfall. I could see a potential water war in the making here.

The groundwater options didn't seem very promising so I moved on to option three: "Tasmania has an abundance of fresh water which could be transported to Sydney should the need arise."

Now which highly paid consultant thought that one up, I wonder?

Think seriously for a moment about the volumes of water required. How many super tankers a day could Tasmania despatch? It sounds about as practical as towing an iceberg from Antarctica. Actually, now that the Southern Polar Ice Cap is melting, it may be a suggestion worth revisiting. Imagine the scene – a giant iceblock coming through Sydney Heads pulled by four destroyers. What happens when it finally makes it to Circular Quay? How do we share it out?

You see there's always bound to be a difficulty with distribution when it comes to icebergs. A super-tanker from Tasmania would have the same problem. I think we could safely put option three into the "clutching at straws" category and progress to option four.

Under the heading "Temporary Water Pressure Reduction", The Plan suggests that: "Sydney Water has the ability to temporarily reduce water pressure in the system and hence reduce the rate of flow to customer's taps. A reduction in flow rate should lead to a reduction in water usage."

Apparently this part of the plan is already being implemented.

"Water pressure will be progressively reduced in much of the metropolitan area over the next few years as part of a cut-price strategy to stem widespread leaks in the ageing mains network," wrote *The Daily Telegraph*'s Kelvin Bissett (21 October, 2003).

This secret strategy was revealed only after the release of an internal Sydney Water report obtained under the Freedom of Information Act which revealed that 10–13% of the city's water supply was leaking away.

Sydney Water's Assets Management General Manager, Paul Freeman, indicated that there may be some slight inconvenience to consumers: "[Residents] might notice it at the tap – it might not come out as fast," he conceded.

If you've ever tried to bathe under a dribbling shower in a Third World country, you'll understand what Paul's talking about here.

Now this might be a fine strategy to reduce consumption, but what if there is no more water left to come out of the tap? Bear in mind this is supposed to be option four of a carefully crafted contingency plan to provide a major metropolis with some degree of water security.

In that circumstance it is a great disappointment.

That leaves us with option five: water rationing.

"It is also possible for Sydney Water to control the volume of water that is supplied to customers." It goes on to say that the corporation "would not undertake such a task unless every option had been tried and failed".

Having reviewed the previous four options, I'd have to say that rationing is the most likely proposition, but theoretically it would not come into effect until dam levels fell to 30%.

That's the essence of The Plan that the highly paid consultants and managers of Sydney Water considered adequate.

But there was another component of The Plan and this was the Sydney Catchment Authority's contribution. I won't bore you with the details of this disgraceful and appaling document, obtained under the Freedom of Information Act, but I would like to draw your attention to its more interesting observations.

The SCA's version of The Plan differs from Sydney Water's in several important details. You would imagine that there would be some level of effective communication between the people controlling the dams and water supplies, in this case the SCA, and those responsible for distributing the product, namely Sydney Water. You'd have every reason to hope that on such important matters they'd get together over a cup of coffee and hammer out an effective strategy.

But, alas, you'd be disappointed.

The SCA version of The Plan says that in 2001, residential use was approximately 56% of the total consumption, with non-residential use being responsible for 26%. Approximately **18% of water was unaccounted for**.

A note attached to The Plan addressed to me stated that Sydney Water wished to point out that "the amount of unaccounted for water is incorrect". This is an unacceptably high and controversial level of waste or loss, far above the official Sydney Water figure of 10–13%.

Now if this was incorrect, why was it not corrected at the time? What is the point of developing an emergency strategy around basic data which is"incorrect"?

The other possibility is that Sydney Water's **official** figure of 10–13% was incorrect. Has Sydney Water been misleading the public about its efficiency levels, or is the SCA incapable of preparing and researching its own Drought Management Plan with any degree of accuracy? These experts are paid millions of dollars annually for their expertise. What are they doing?

Another major inaccuracy is the fantasy of a giant desalination plant that was theoretically supposed to come into the picture when the total storage reached 25%, the equivalent of one year's supply.

> "The Drought Executive Committee will initiate a tendering process for the construction of one or more desalination plants. At this stage the construction of one or more desalination plants will take approximately 20 months."

So with one year's supply left, the SCA proposed to begin a process that it optimistically estimated would take one year and eight months to complete.

As we have seen in a previous chapter, *no* large desalination plant has been completed on schedule, but even if this were possible, that plan would still leave the city without any obvious source of water for eight long months.

This is a very serious discrepancy. As any health professional will tell you, humans cannot go without water for more than seven days.

It gets worse.

Reading on – the whole SCA plan only runs to 19 pages – we finally discover that **the construction cost of a plant with a 500 ML/day capacity is estimated at $750 million**.

The largest and most modern seawater desalination plant in the USA was supposed to provide 125 ML/day. It was supposed to take four years to build at a cost of US$110 million. It is now two years behind schedule and still not working properly. Its final cost is not yet known.

The SCA proposal envisages a hypothetical model that has never yet been built, and they hope to construct it in 20 months?

In the final analysis, the SCA plan, which begins with a high level of concern that "the reservoirs could empty in extremely severe droughts, potentially leaving the Sydney Metropolitan area and surrounds without water", ends up with a ridiculous strategy that is unworkable – one that appears to have developed in someone's fertile imagination. To use the SCA's own words: "That scenario is totally unacceptable."

And so is the rest of the SCA emergency plan, which includes familiar feel-good, oft-repeated themes such as "various effluent reuse schemes and rainwater tanks".

It does, however, contain some important information, including the statement: **"Groundwater is not a viable alternative source as current resources in the wider Sydney region are negligible."**

The advice appears to have been ignored by the NSW government who, in 2005, allocated $4 million dollars to "explore groundwater options".

On 23 August, 2004, Minister Sartor's office issued a statement denying that these plans we've been examining were in fact, government policy.

That's true enough: before July 2004 there was *no* government plan or policy to deal with Sydney's water crisis, other than the two previously mentioned. If such a plan existed it has yet to see the light of day.

In October 2004, with much fanfare, the state government finally announced that it had in fact developed a strategy that would meet the challenge and secure Sydney's water future.

The Metropolitan Water Plan (www.dipnr.nsw.gov.au) contains many of the usual promises and future possibilities (desalination, recycling and groundwater), but these are still only at the feasibility stage that they should have been at 10 years ago.

Two additional options have been introduced, both identified previously by the SCA as contingency alternatives in extreme drought situations. The Government's media unit tried to present these in the best possible light.

The best way to disguise a desperate measure is to present it as a bold new initiative. One example is the proposal to "access deep water at the bottom of dams". Why hasn't anyone thought of this before, you might ask?

The water below the outlet pipe at the bottom of a dam is called "dead" water because that's where all the debris, junk and toxic contaminants accumulate in the mud and sediment.

Sediments are composed of fine particles of dirt and sand. Contaminants and pollutants attach themselves to these particles and are carried to the bottom of rivers, lakes and dams. Sediments can store, concentrate, and release contaminants. Both high and low flow conditions favour the release of contaminants that may have been stored for years.

Sediments in the bed of the Brisbane River were found to be so contaminated with DDT, deildrin and other poisonous chemicals that they were classified as toxic waste. DDT was banned in Queensland in 1987 – but it remains stored in the mud at the bottom of the river.

The sediments on the bottom of our dams are similarly contaminated. Accessing the dead water could well stir up these particle layers best left undisturbed.

Let's look at the example of a rainwater tank. You will see that the tap and the outlet pipe are not right at the base of the tank but some distance above it. Any rubbish or debris or dead birds that might fall into the tank will sink, so it's wise to lift the outlet high enough to avoid drinking the muck that might end up on the bottom.

Dams operate on the same principle: all the muck collects below the level of the outlet. This water represents about five per cent of the total dam volume.

The SCA Drought Management Plan warns that this is not an ideal source of drinking water.

> The remaining 5% of the storage refers to unusable storage referred to as a "dead" storage. This may be accessed if the usable storage of the reservoirs is fully depleted, however the quality of dead water is considered poor.

But in a multi-million dollar advertising blitz on prime time television during March and April 2005, the Water for Life campaign showed the dead – now called "deep" – water as a new, previously undiscovered source of "quality" water.

Like Lazarus, the dead water has miraculously been resurrected and given a new lease of life.

Dead Water for Life – check it out for yourself at www.waterforlife. nsw.gov.au/government/dam_water.shtm

Accessing this dead water will cost around $106 million and give Sydney six month's additional supply. That's about $17 million a month for a plan that will mix contaminated water with what remains in the already depleted dams. This will result in a diminished quality of water both now and in the future, water that will require increasing levels of chlorine disinfection to make it "safe" for human consumption.

I, for one, will not be drinking it.

Another desperate option is the proposal to take more water for Sydney from other more distant communities, specifically the Shoalhaven River, which supplies the Southern Highlands. This proposal is scheduled for "community consultation" with "key stakeholders". It will not be well received.

Planning is a process that needs time, resources and commitment to the process. The foregoing critique of the NSW government's inadequate and last-minute planning for the country's largest metropolis is offered as a case study for those who may be interested in the city's future.

To summarise the position as of April, 2005.

Sydney's dams are at record lows. Although heavy rain has been falling on the coast, little has fallen on the catchment. On May 10, 2005, the city of Goulburn, for instance, was down to 12% of its drinking water supply, while inflows to Sydney's storages have dropped by half since 1990.

For the past 14 years the only government response to this looming crisis has been to urge Sydneysiders to limit their consumption. This has been very successful. Consumers responded positively, reducing demand to unheard-of levels. But no other plan for the future was implemented. The authorities hoped for rain. It has not arrived.

In July/August 2004, the government started to express concern.

By then the Premier had begun to focus on climate change. In October 2004, in The Metropolitan Water Plan, Mr Carr revealed that:

> International and Australian scientific experts have told me that the greenhouse effect is already having an impact on our climate and could even be influencing our current drought. That is why ... we are thinking about this problem now ...

It's kind of late to start thinking about the problem long after its effects are manifestly obvious to the rest of the population.

A plan is something developed carefully and thoughtfully over a period of years in order to anticipate a problem. And it's not that this government was incapable of making plans. Sydney has plenty of other well-documented plans – plans to grow its industries and expand its suburbs, to encourage development and build vast tunnels and freeways – but water, the essential ingredient in all of this, has been neglected.

More than a billion dollars has been taken in dividends from Sydney Water and the Sydney Catchment Authority, money which might well have been spent on recycling schemes or replacing the decaying mains that allow between 10 and 18% of Sydney's precious water to go to waste.

Unfortunately, this policy vacuum is not limited to NSW or indeed to Australian water authorities as a whole.

In his well-researched overview of the global water situation, *Every Drop For Sale – Our Desperate Battle Over Water in a World About to Run Out*, Jeffrey Rothfeder observes:

Governments, in wealthy and poor countries alike, have mishandled water supplies so badly for such a long time that their mistakes, poor planning, and ill-conceived policies are a prime reason that the amount of available water today is dwindling. Almost everywhere, the water infrastructure – the sewage pipes, the pumps, and the quality control stations – is falling apart, mostly out of neglect, creating pools of wasted water.[1]

Sydney's leaking mains are an ongoing scandal. During 2002 and 2003, Sydney Water admitted to losing at least 10% of its water supply due to leaking and broken mains pipelines.

That's twice the national average.

Another 2.5% was allegedly stolen, or lost through inaccurate metering. In total, that amounts to 85,000 megalitres, or enough to supply the city for almost two months.

Aging and decrepit mains are a continuing drain on finance and water quality, according to a Sydney Water discussion paper of August 2000.

Older cast iron and galvanised iron pipes are prone to more corrosion and may have unlined fittings. These types of pipes are prone to cause dirty water and harbour micro-organisms....

Roughly 75% of mains are cast iron with cement lining ... Nearly 30% of water mains are over 50 years old. In some areas, over 80% are over 50 years old.[2]

If the government's figures are correct, around 60 kilometres of mains are replaced every year at an approximate cost of $500,000 per kilometre. At that rate it would take 300 years to replace the system, at a present-day replacement cost of $10 billion. Since the optimum life of a mains pipe is less than 100 years, you can see that this is a costly business. But as costs increase every year, delay or postponement creates an added burden.

And that's just the water mains.

Then there's the sewerage system that is in a similar state of disrepair. Spending billions repairing pipes is ultimately essential but because it's not very glamorous or appealing, politicians prefer to spend money on things like wars and operas and Olympic Games and overseas study tours to Europe.

No wonder a 2001 report from the Institution of Engineers rated the condition of the entire nation's water and sewerage networks as "relatively poor".[3]

And guess who owns these decaying assets?

You do.

Expect a bill for repairs in the near future in the form of increased taxes, rates and charges.

The performance of the Board of Sydney Water led by Ms Gabrielle Kibble (Madame Borgia as she is known to her many detractors) has been a complete and utter disgrace. Incompetent business-oriented managing directors of Sydney Water have resigned or been sacked one after another when failed "initiatives" wasted millions of dollars.

There is no sense of united purpose between the SCA, Sydney Water and the three ministerial portfolios that have joint responsibility for the supply.

A headless chook would have a better sense of direction than this lacklustre group.

In May 2005 Bob Sendt, the NSW Auditor-General warned that the Government's Metropolitan Water Strategy had serious deficiencies. There were no built-in contingencies to allow for climate change or for the impact of a greater than expected rise in Sydney's population. "I am not saying the plan should consider a doomsday scenario but it should build in a buffer," he said (*The Sydney Morning Herald*, 5 May, 2005).

Sooner or later, the NSW and other state governments will wake up to the fact that water is central to our survival and acknowledge that by creating a ministerial portfolio dedicated solely to water, administered not by political cronies but by people with experience, people who know what they are doing.

None of the plans currently proposed offer more than a brief reprieve.

If rain does not fall on its catchments during the next two years, Sydney could well be the world's first capital city to run out of drinking water.

Let's hope that nightmare never becomes a reality.

Liquid Assets
WHO OWNS THE RAIN?

We are currently facing a global water crisis, which promises to get worse over the next few decades. And as the crisis deepens, new efforts to redefine water rights are under way. The globalised economy is shifting the definition of water from common property to private good, to be extracted and traded freely. The global economic order calls for the removal of all limits on the regulation of water use, and the establishment of water markets. Proponents of free water trade view private property rights as the only alternative to state ownership and free markets as the only substitute to bureaucratic regulation of water resources.

– Dr Vandana Shiva, *Water Wars: Privatization, Pollution and Profit*, 2002.

"Water is already a huge business and it's going to get bigger and bigger around the world. The provision of water supplies ... will boom over the next decade. Those companies that position themselves well will reap the benefits."

– Richard Pratt, *Business Review Weekly*, 9 October, 2003

Nothing in life is free.

– John Howard, 29 April, 2005

IN BOB DYLAN'S PROPHETIC song "Well, Well, Well", the mythic man who stole the water is the epitome of evil, condemned to swim forever in eternal darkness. In today's brave new world such men are often hailed as visionary leaders.

Unfortunately there's more than one – in fact there's a whole cabal of conspirators planning to steal *your* water. And they're not going about it in a sly, underhand fashion as you might expect – no, it's all happening out there in the bright glare of the media spotlight.

We are at an important point in our history where significant choices are being made – choices with serious implications for the nation's future.

Like any artful confidence scam, it begins with a story that sets the scene and gives the conmen the appearance of legitimacy. In its simplest form, the story goes like this:

Australia is the driest inhabited continent on Earth, so water is a scarce commodity. In the past, rural Australians have not used water carefully or wisely. In order to make people more careful, it is necessary – critical even – to take this water, which up until now has been public property held in common, and alter its legal status.

In 1994, The Council of Australian Governments (COAG) decided that water should be recognised as a separate property right for farmers. This was a primary part of water reforms aimed to improve the health of rivers and improve the efficiency of rural water use.

Now, instead of being the nation's common property, water will be bought, sold and traded just like any other commodity, in order to allow "market forces" to determine its true value and make sure that it is used appropriately. This not only helps the environment, it is an essential element in a strategy of "wealth creation" which will ensure the future prosperity of this great nation of ours ... and so on and so forth.

You must know by now that when men of vision in high places start to talk earnestly about "wealth creation" and "the future prosperity of the nation", there are always bound to be winners and losers.

Think Alan Bond, John Elliott, Rodney Adler. Think IT geniuses and dot com millionaires. Think Christopher Skase and his aptly named Mirage Group.

"Wealth creation" is like lotto – for every winner there are millions of losers.

So who might the winners and losers be when it comes to water reform? In every reform so far, the Australian people have always been the big losers. No prize for guessing that one correctly.

And the winners?

They're the usual suspects, the same crowd who've always been major beneficiaries of government largesse – big agribusiness corporations, millionaire entrepreneurs and speculators, and the banking industry who know a good profit-making venture when they see one.

As I mentioned previously, the rationale for this massive movement of our common property from the public to the private sector is called reform. In my opinion, the reform agenda has been the single most destructive force in recent history.

It has disrupted the lives of millions of Australians while simultaneously robbing us of our birthright, eroding our independence and destroying many of the nation's institutions that have previously supported our way of life.

And all for what?

How have we benefited from the raft of changes that have been imposed on us since the Council of Australian Governments adopted the reform agenda in 1994?

Think about it. Cast your mind back. Have things changed for the better during the past 10 years?

Remember when health and education were free, when there were no queues at banks or hospitals, when the phone bill was not a major item of household expenditure, and you didn't have to work overtime just to break even.

Take tax reform for example.

We were promised changes that would simplify the tax system and make it more equitable. Instead the goods and services tax drove many small businesses to the wall and tripled the cost of bookkeeping for everyone else. It raised the cost of living by 10% – and still we pay record rates of income tax. Call up the tax department to complain and you won't even find a human being on the other end of the phone.

Then there was telecommunication reform.

Rates and charges rose rapidly while the quality of service declined. Thousands of employees were consigned to the scrap heap by over-paid, ego-driven competition freaks anxious to improve the budget bottom line. The sale of Telstra saw another big public utility pass into private hands. A lot of wealth was created. Now instead of a human voice, there is a sympathetic robot on the line.

Banking reform meant record profits for shareholders as thousands of branches were shut down and staff made redundant. Customers are offered the "convenience" of the Internet or the autobank, or the inconvenience of waiting like sheep in roped-off queues in order to talk to a live person. Try calling their number and yep – there's yet another mechanical friend waiting to help you. All you need to do is press the button for the option you require.

Here in NSW we've had a few years of inspired transport reform to reflect on. Large bus companies are going bankrupt while the performance of the rail network has reached such an all-time low that it seriously threatens the Carr Government's chances of re-election. Ring CityRail and tell the nice robot how you feel about it.

Reform of the NSW health system led to ward closures, hospital budget shortfalls, long queues for elective surgery, ambulances being turned away from overworked emergency units, staff shortages and increases in medical misadventure, unnecessary injury and death.

I think it's fair to say that most Australians have had enough reform and structural change to last them a lifetime.

Once, we were proud of the fact that anyone with ability was entitled to a free university education regardless of their financial situation.

Reform changed all that.

Now we have a generation who've had to mortgage their future in order to pay for something we once regarded as a right. That's yet another example of how governments can create wealth for themselves by taking something away from you and then selling it back at market prices.

If we believe that our water should remain common property, then now is the time to assert it. Otherwise it will go the same way as free universal healthcare, education and welfare. Before long, even the old-age pension will be a thing of the past.

The process of preparing the public for water reform includes "consultations" and "forums" where we learn that we have been having a free ride; we have not been paying for the water we've used. This is true. Our water bills simply reflect the cost of collection, treatment and distribution. The water itself was free, because it was held in common and owned by no one.

This principle has its beginnings in the nineteenth century when in order to facilitate the distribution of water in the colonies, legislation was introduced which gave the Crown the right to use and control water. When Australia became a federation, this right remained in the hands of the states.

Land ownership was also vested in the Crown and theoretically held in common until the 1860s, when an Australian surveyor developed a property title concept known as the Torrens system. This remains the foundation of property ownership today.

The Torrens system precipitated a massive land grab by squatters and graziers, during which a vast area of the country passed from public into private ownership. Those who converted their Crown leases into freehold titles became wealthy landowners overnight.

Of course, nobody gave the land rights of the indigenous occupiers a second thought.

In his *Devil's Dictionary*, social satirist Ambrose Bierce describes aborigines as "persons of little worth found cumbering the soil of a newly discovered country".

This certainly reflected the prevailing attitude in the early to mid-nineteenth century. Here was this whole big empty continent with just a few Aborigines, who'd done nothing to improve or cultivate the land, standing in the way of progress, civilisation and wealth creation.

Winners and losers again.

White wins, black loses. Place your bets again folks!

The Federal Government's National Water Initiative now proposes to take the water that once belonged to the indigenous inhabitants (if indeed it belonged to anyone), and which now theoretically belongs to us all, and give it to private investors and existing users so that they can buy, sell, trade, mortgage, do whatever they want with it.

Just like that.

No public participation, no public input and, above all, no public transparency.

The bill to set up the National Water Commission, which will oversee the "Initiative", contains a secrecy clause that prevents the Commission from releasing any information about the state of Australia's water supplies without ministerial approval, according to a report in *The Advertiser*, 3 December, 2004.

So who gets how much of the nation's water, and how much they paid for it, suddenly becomes secret politician's business?

We are talking about the movement of billions of dollars of public money here, in the hands of a Government that thinks a matchbox is a weapon of mass destruction, a Government that took this country to an illegal and immoral war based on a lie, a Government that has consistently misled and deceived the Australian people.

Trust us, they say.

During the 10 years since water reform began, Australia's rivers have been in a state of continuous decline. Water reform has failed in its stated aim of improving the condition of our aquatic environment.

According to the *Australia State of the Environment 2001* report, 26% of Australia's surface water management areas are close to, or have exceeded sustainable limits. This includes the supplies of all capital cities except Darwin and Hobart.

The report continues:

> Many river systems in the Murray-Darling and along the east coast of Australia are either overdeveloped or approach full development status ...

> ... pressures on inland waters have increased and many water bodies in the developed southern and eastern areas of Australia are significantly degraded as a result of activities in the catchments and water extraction for agriculture.[1]

In the case of groundwater the news is even worse.

> Groundwater available for allocation has reduced substantially in the last decade and is now overused and over-allocated ...

> ... some groundwater resources are already overdeveloped as the rate of extraction exceeds the rate of recharge. These include the Great Artesian Basin, many small aquifers in the Murray-Darling Basin, the Perth basin, and aquifers along the east coast of Australia.[2]

In conclusion, the report pointed out that:

> Although the National Water Reforms Framework include provisions for groundwater, groundwater reform is lagging behind surface water reform in most states and territories.[3]

The drought that has affected much of Australia for the past eight years has seen the quality and quantity of our water supplies deteriorate even further than the *State of the Environment* report indicates.

The predictions from the CSIRO and other climate change experts are that there will be progressively reduced rainfall over much of Australia in the years to come. That means less water for everybody, except for some lucky folk up north.

In the past, irrigated agriculture has accounted for 70 to 75% of the nation's water consumption. This water is currently allocated by the states using a system of licenses.

The National Water Initiative proposes to convert these water licences into perpetual water "rights". It is part of the legal process of separating the land from the water and having a separate title to each, thus making water a commodity in its own right. Is this a good idea in a time of increasing scarcity and resource depletion?

One of the factors complicating the current water crisis is the passive public acceptance of the increasing commodification of water.

Does that matter?

Let's look at the implications and philosophical direction of this concept, and follow it to its logical conclusion.

Who, for instance, owns the rain?

This is central to the issue.

Who owns the rain?

In principle we do. Theoretically, Australians own all the rain that falls in Australia – the rain that becomes the lakes, rivers and streams that flow over the land and the groundwater that flows beneath it.

That's our water. It's our common property, our birthright if you like.

Okay. Now suppose you collect the rain that falls on *your* roof and store it in *your* water tank. That water becomes in a sense your property, to use as you want to.

Simple.

But what if you had a gigantic roof – as big as the Sydney Opera House? You'd have ten or twenty times more rain than a suburban house, stored away in your massive tank.

Now, suppose you had a farm with sloping hills, and you put in some dams to catch the rain that fell on *your* property. There are several million such dams in Australia, 300,000 in Victoria alone. By extension that's *their* water, just as the water that falls on your roof is yours.

But that's also a lot of water that used to flow into our rivers and streams, water that's withdrawn from the common store.

Now think about a larger property with an enormous private dam on it, 200,000 hectares in size, big enough to hold 500,000 million litres – the equivalent of Sydney Harbour. That's the size of Cubbie Dam, Australia's largest private water storage located near Dirranbandi in South-east Queensland.

Cubbie is only one of many such private dams in the Condamine/ Balonne River catchment. These dams are capable of holding back the equivalent of the entire average annual flow of the river.

The owners say that's *their* water, to use, sell, trade or pollute.

And how much does Cubbie Station pay for all that water they collect? The sum of $3,700 per year, about the cost of a medium-size suburban rainwater tank.

And what is it worth at the current market prices paid by some of the other irrigators? Around $15 billion.

There are obvious differences between collecting the water from your roof and Cubbie Station's massive storages. The first is for personal use – the other is for personal gain. Common resources are always threatened when greedy people take more than their share.

Winners and losers again.

Business wins, the environment loses.

And when the environment loses, ultimately *you* lose.

Let's look at another example – bottled water.

Most irrigators and water bottlers pay between 1 cent and .05 of a cent for 1000 litres of ground or surface water. Urban water utilities pay the same rate.

By the time it reaches the domestic consumer, a kilolitre of Sydney tap water costs 98 cents.

Put the same water into 500 mL plastic bottles with fancy labels, and the consumer cost rises to $4000 – and sales are increasing every year.

From .05 cents to $4000 is a handsome rate of return, even allowing for the costs associated with pumping, bottling, marketing and distribution.

The profits go to shareholders, while the costs to the environment – depleted aquifers, diminished river flows, increased acidity and subsequent contamination – are borne by the entire community.

That's why investors are getting into the water business. When water becomes a commodity, individuals profit at public expense.

To return to that rain we thought we all owned in common: Remember the old song that goes: "Every time it rains, It rains pennies from heaven ..."

Imagine that the rain, literally and figuratively, *is* pennies from heaven, and that the sum of those pennies is the capital and assets of our own national water bank. We need to ask some basic business questions to find out how our bank has been performing over the years. How does it measure up against the performance of some of Australia's other banks, such as Macquarie, or the National or Westpac?

Unlike real banks, our national water bank gives most of its operating capital away for free. The costs of the 447 publicly owned dams and infrastructure in Australia were essentially losses, because these facilities were constructed at the taxpayers' expense.

So, according to an economic rationalist perspective, our hypothetical water bank never had a chance. It lost billions from the very beginning.

The successive governments responsible for managing and distributing our water for the past fifty years have failed to live up to their responsibilities. And the CSIRO and the army of water scientists and hydrologists, the experts who advised those governments, have been at the very least, complicit in that mismanagement.

Instead of enhancing and protecting the water that represents our most valuable natural capital, they have given it away, let it be abused, neglected and wasted. By over-allocating it they even gave away water that didn't exist.

The over-allocation of water has created a rift in rural Australia. Some drought stricken communities are fighting for what they were led to believe was their rightful allocation – water that did not in fact exist and can never be fully supplied.

Who could blame them for their anger?

As John Grabbie, manager of a large irrigated cotton farm on the Condamine–Balonne, told Ticky Fullerton on the ABC's *Four Corners* program in 2001: "If the river is being destroyed, it is destroyed because irrigators are doing what they were told to [by government]."[4]

It is true that governments have implemented unwise and ill-conceived policies that sanctioned and promoted unsustainable use and deliberate over-allocation. The negative social, economic and environmental consequences of these decisions are still emerging.

In this instance, it is science that has failed us.

Was our faith in science misplaced? Or was it the scientific community who let us down?

In 2002, the popular radio broadcaster Alan Jones started a public appeal for funds to help drought-stricken rural Australia – The Farmhand Foundation. This generated a flood of fresh ideas and concepts, some practical, some not. They were all compiled into a large and useful volume called *Talking Water – An Australian Guidebook for the 21st Century*, published privately in 2004. (Download a free copy at www.farmhand.org.au).

Instead of listening to the voice of the public and joining in the debate that Alan Jones initiated, some members of the scientific community chose instead to construct their own agenda.

On Thursday, 10 October 2002, eleven of these luminaries met for a meal at Sydney's upmarket Wentworth Hotel, and after dessert they decided, as they put it, to "take on" Alan Jones by combining their expertise and proposing their own solution.

"This was a sobering challenge," Professor Peter Cullen told a CSIRO Land and Water seminar in Adelaide on 22 March, 2004, and the experts rose to meet it.

"It was not difficult," Cullen continued, "and in 15 minutes we had agreed [sic] the five-point plan that was the foundation of the Wentworth Group's *Blueprint for a Living Continent*." (Read the latest version for yourself at www.wwf.org.au)

Fifteen minutes was all it took these guys to solve the nation's water crisis. Wow!

But, in their rush to embrace the Wentworth Group's vision, the media overlooked one important fact.

The eleven prominent scientists who proposed this solution have been the advisors to governments, state and federal, and a driving force behind government policy for the past fifteen years. The current water catastrophe unfolded during the very time that they were at the wheel, funded with millions of dollars of tax-payers' money.

Why did they wait for so long before speaking up? What on earth have they been doing for the past fifteen years?

Professor Cullen, the founding director of the Canberra-based Co-operative Research Centre for Freshwater Ecology (CRCFE), has been involved in freshwater research for 30 years. In 2001, Professor Cullen received the Prime Minister's "Environmentalist of the Year Award" for his "outstanding contribution to improving the performance of governments and communities in the conservation and management of water resources".

I may be short sighted but I fail to see any improvements by governments or communities in the conservation and management of water resources.

At the risk of being repetitious, all of Australia's environmental indicators point to a consistent decline during the past 20 years.

The only factor showing upward trends has been the amount of money wasted on unproductive and worthless science which obsesses about minute parts of the problem without addressing the whole.

Perhaps the limitations of science have not been explained to us lay-people.

According to Professor Cullen:

> ...all science can do and should do is to identify what the consequences of various decisions will be, hopefully identifying long-term as well as short-term changes. (*Watershed*, June 2001)

So scientists are *not* really active participants but passive voyeurs, assiduously monitoring the gradual decline of the environment at our expense. It's not their job to advocate, they say – not their responsibility to speak up when they see ignorant, stupid people in positions of power acting contrary to the national interest.

"Separating scientific advice from advocacy is critical," Professor Cullen's successor, Gary Jones, told his staff at the CRCFE in *Watershed*, the Centre's monthly newsletter. Professor Jones continued:

> ... we have learnt that it is very important to separate scientific advice from environmental advocacy. It is all too easy for environmental

scientists, even the very best environmental scientists, to embed their views (albeit unintentionally) about desirable ecological outcomes in the advice they are providing to government and the public. In doing so there is a very significant risk of alienating key stakeholder groups. (*Watershed*, December 2003)

Alienating key stakeholders, such as the water industry and the Federal Government who jointly fund the CRCFE, seems to be a major consideration. After all, they're the paying customers.

The CRC exists to generate new knowledge from its research programs, and to deliver that knowledge to the water industry to help them manage our precious water resources. **The industry provides significant financial resources to help the CRC meet this aim** [my emphasis]. (*Watershed*, March 2003)

And what if the interests of the water industry and the environment conflict?

It is not the role of the scientific community to decide on the compromise between the competing values of production and the natural environment. Rather it is for the community, through the political process to determine this compromise. (*Watershed*, February 2003)

So ultimately the buck is passed back to the community.

It's not our problem, the scientists are saying. *You* elect the politicians, and if *they* choose to make bad decisions in the absence of any scientific environmental advocacy, it's *your* problem, not ours.

And all along I imagined in my naivety that science was working for the public good, that these guys were advocating on behalf of the environment and our natural capital.

In order to reverse – or even to slow – the rate of environmental degradation, unpopular and difficult choices need to be made, choices which will affect the livelihoods of many. Since politicians appear to be more concerned with their short-term political survival than the long-term survival of the planet, those choices are often put in the too-hard basket.

It is unfortunate that the cycles of nature do not harmonise with the cycles of politics.

So in the end, who actually makes these difficult choices about the fate of the nation's water supply?

Politicians? Scientists? The public?

Or someone else? Could it be, as businessman Richard Pratt suggests (and he should know), that:

> It's not the leader of the party that's the leader of the country ...
> I'm looking at a group of people beyond the politician. These are the
> leaders of our country, maybe the hidden face of the establishment,
> they are the leaders and they have to be the leaders of the country.[5]

Interesting isn't it?

Here we are fighting for freedom and democracy in Iraq while our own cardboard cut-out pollies are manipulated and controlled by "the hidden face of the establishment".

Australia's second-richest man seems to have eclipsed the Wentworth Group of Concerned Scientists when it comes to political influence. His vision is as wide as his wealth is vast, and there's something about money that seems to unblock the ears of otherwise deaf politicians.

As the old saying goes, money talks... often very loudly.

"Pratt plans takeover of the whole dang NSW water system and the National Water Initiative, too", read the headline of *The Water Report*, a leading industry newsletter on 24 January, 2005.

The article continued:

> After gaining about $13 million from the federal and NSW
> governments to pay for a feasibility study, two versions of a final
> Pratt plan proposed 17 NSW policy changes alone. It appeared the
> idea was the state would change the rules to guarantee the income for
> his new enterprise Pratt Water through a Private Public Partnership.
> The apparent plan was the sale of operations/assets to Pratt.

As I pointed out earlier when men of vision embark on an exercise in wealth creation there will always be winners and losers.

The Water Report again:

Winners in the Pratt plan:

- Pratt via sales of a new type of pipe;

- Pratt through the partnership with ANZ bank (where Pratt/ANZ would lend money to irrigators to buy Pratt pipes); and

- Large irrigators (such as the NSW Rice Growers Cooperative), which would trade "saved" water in a water options market

- Pratt (as National Bonds would then fund non-NSW irrigators to buy Pratt pipes)

- Pratt's plastic dams would store and trade water.

(For further details – see www.prattwater.com.au)

Central to Pratt's vision is the idea of saving water by reducing the massive losses due to evaporation and seepage in irrigation water channels. This loss is indeed impressive.

Australia has about 70,000 kilometres of open water conduits of one sort or another. These include stock and domestic supply systems, surface drains and bore drains.

The amount of water lost in irrigation channels each year is twice the amount used in all of Australia's towns and cities put together.

The three main factors responsible for this massive loss are, in order of magnitude, evaporation, leakage (where water flows onto adjacent land) and seepage, where water soaks into the ground beneath the channel.

Richard Pratt proposes to pipe all of Australia's irrigation channels with a collapsible woven plastic piping that can be laid along existing canals and trenches. If this is combined with drip irrigation and other state-of-the-art irrigation technologies, water savings of between 40 and 50% are possible.

This is not a new idea. Similar projects are under way in other parts of the world. But hydrologists are discovering that there are hidden environmental costs associated with the practice.

An editorial in *New Scientist* (21 August, 2004 – see www.newscientist .com) drew this issue to international attention:

> Hydrologists have till now comforted themselves with the belief that we are so inefficient in the way we use water, particularly for irrigation, that modest investment could transform the situation ...
>
> More than two-thirds of the water we grab from nature is intended for irrigating the crops that feed the world, but nearly two-thirds of that never reaches the plants. Instead water leaks from distribution canals and percolates underground or it evaporates from flooded fields. Capture that wasted water and everything will be OK.

But it is not that simple.

New Scientist points out that research from 90 institutes around the world showed that much of this theoretically wasted water recharges groundwater supplies that are often themselves the base flows of rivers. So more efficient technologies upstream may in fact lead to reduced river flows downstream, as well as lowered groundwater tables.

This has already occurred in Sri Lanka and northern Mexico.

The Pratt Water Murrumbidgee Project acknowledges that: "the adoption of water efficient technologies and practices will result in lower accessions into groundwater ... This could result in watertables dropping by up to 2 metres ... ground salinity levels may also rise ..." (*The Water Report*, 24 January, 2005).

So there are risks and benefits in the Pratt plan: Potential risks to the environment versus guaranteed benefits to investors.

The cost of the project exceeds $10 billion.

The ten billion dollar question is who will pay for it, and how?

The Pratt Water sales pitch comes in a lavishly produced volume called *The Business of Saving Water*. It is thoroughly researched, and well designed and presented. The book is certainly impressive.

And that's what it's meant to do – to impress planners and politicians.

I have no doubt that many are contemplating Pratt's eloquent and comprehensive proposals which offer a tempting face-saving alternative in the current policy vacuum.

One of the more interesting parts of the Plan is that water "saved" by investment in the pipelines and such becomes the property of the investor who put up the money.

The Business of Saving Water put it this way:

> Saved water should be recognised as a new category of water,
> with secure title commensurate with the investment effort
> taken to secure it.
>
> Private sector investment in water efficiency must thus be
> accompanied by a clear and unequivocal property right to
> the water saved. This right must take precedence over
> existing water security levels to provide more guaranteed
> access to the saved water during times of scarcity.[6]

This may be an advantage for individual investors, but it's difficult to see how the environment would benefit. The CSIRO's Dr Wayne Meyer also expressed doubts, telling an Irrigation Australia conference in 2000:

> The notion that improved water efficiency will free water for
> environmental purposes is simplistic and will not work. On
> the contrary, the demand for irrigation use is likely to increase
> with water savings unless there is some explicit mechanism to
> encourage saved water to be reallocated.[7]

In addition, the billions of dollars in investment, and subsequent returns on that investment, will inflate the value of water and increase the cost of production and, ultimately, the cost of food and other staples. This cost will be passed onto the community.

While Mr Pratt's research is commendable, the transfer of public water supplies to private investors could well complicate an already precarious situation.

The issue of water rights is very complex. It has produced mountains of paper. The Productivity Commission's report alone runs to 320 pages. Politicians, economists, and local and global investors are using this complexity as a screen to conceal the transfer of our last remaining common asset from public to private property.

And they say they're doing this to protect the environment!

In an interview with *The Sydney Morning Herald* (1 March, 2004), Craig Knowles, NSW Minister for Natural Resources, began by saying that a national approach offered "the best chance to replenish NSW's rivers".

An innocent bystander could be forgiven for observing that the best way to replenish the state's rivers might be to let their natural flow reassert itself.

But nothing is as simple as it seems. The article continues:

> As part of that commitment, Mr Knowles has announced **perpetual** [my emphasis] water licences in order to give farmers and irrigators greater economic security.
>
> Mr Knowles said private investment was needed to ensure enough water remains in NSW's rivers to guarantee their health.
>
> **"Without the active consideration of private involvement there would not be any water for the environment** [my emphasis]," he said.

That's the plan?

The nation's rivers are going to be replenished by giving their water away in perpetuity or selling it on the open market, so that the taxpayers have to buy the same water back for the environment, or for urban consumption?

The bottom line is this. If we passively allow the commodification of our last common resource, what was once an asset will suddenly become a liability.

Who will buy the water for the environment from the investors who own it?

We are on the verge of giving away something rare and precious and then having to buy it back at market prices so that it can fulfil its original purpose.

This is not a solution.

This is not a magic bullet that will suddenly solve the problems created by decades of neglect and mismanagement.

Instead of reversing the rate of depletion and degradation, privatising water through property rights will accelerate it.

Competition for control and ownership will unleash water wars by pitting one person against another, region against region, rich against poor, rural areas against cities.

If we allow this to happen, we may find that we have simply dug ourselves deeper into the hole that we're already in.

Only to find precious little water at the bottom of the pit.

Breathing Space

2005 began, like the year before it, with a looming water crisis.

Sydney's dams were at their lowest levels – 43% – since they were built. Perth was grappling with a long-term decline in rainfall and water availability. The Gold Coast, Sydney, Canberra, Melbourne, Adelaide and Perth and many country towns all had water restrictions. In many parts of Australia, residents will never again be permitted to hose down paths or turn on sprinklers.

 – Asa Wahlquist, *Weekend Australian*, 19 March, 2005

Permanent water restrictions in big cities could offer a "breathing space" as Australians begin to understand and debate the massive changes demanded by the Federal Government's National Water Initiative.

 – Professor Paul Perkins, *The Australian*, 1 March, 2005

BREATHING SPACE – time to rest and reflect.

Given that our water has been grossly mismanaged at all levels and that in spite of the state and federal Government "plans" for our water future, our situation is not going to change for the better in the foreseeable future, what can we do about it?

What can we do?

Where do we begin?

We could begin by reconsidering the fundamental flaws in our vision of the natural world – the ideas and ideals that led us to where we are today. Because ultimately the long-term solution to the water crisis is not about how often we flush the toilet, it is about changing our perspective and recognising that we are not the masters of nature. Anyone who witnessed the power of the 2004 Asian tsunami couldn't help but realise that the awesome energy of the elements is far beyond our control.

We exist as part of a global ecosystem, one that is badly overstressed by centuries of thoughtless abuse and neglect. Those stresses are now beginning to affect our daily lives. The future magnitude of those effects will be in direct proportion to the level of damage suffered by the ecosystem that supports our existence. We need to reduce environmental stress – and to reduce the level of stress in our own lives as well.

So let's think about how we might achieve both goals simultaneously. That's not as difficult as it sounds because the energy that's causing most of our personal, social *and* environmental stress originates from the same source.

It's a concept called "growth".

Now obviously there are many different kinds of growth. Growth is a natural process. Things grow, mature and die – and the cycle begins again. That's one sort of growth.

But we have become enchanted by the vision of another form of growth that is not cyclic but continual and exponential. Some have become so entranced with this vision that they have lost or abandoned their capacity to reason. Growth, whether it be population, industrial or economic, has become an Australian political and social ideal – and more. For many economists and political leaders, growth is such an overwhelming obsession one could almost call it a religious belief.

This is insane. Someone needs to say it loudly and clearly.

Continual growth for its own sake is insane because it is ultimately self-destructive.

Let me explain why.

Exponential growth describes a process of doubling, and doubling yet again, and so on. Linear growth – the pattern with which we are most familiar – occurs when the increase is constant over a given period of time.

For instance, if you put $10 in a jar each week, at the end of the year you will have $520. That is linear growth. It does not depend on the amount already accumulated.

A quantity is said to grow exponentially when its increase is in direct proportion to what is already there.

So in the case of our $10 weekly deposit, suppose that it earns compound interest at the generous rate of 10% calculated weekly (Mafia moneylenders call such extortionate rates of interest "vigorish" or "vig").

At the end of the first week you will have $11, the second week $23.10, the fourth $35.41, the fifth $48.95, the sixth $53.84 and so on. By the twelfth week there would be $172.87 in the bank, as opposed to $120 without any interest.

It is obvious then that growth which multiplies upon itself will produce substantial increases very quickly.

Here's another example.

A Persian legend tells about a clever courtier who presented a beautiful chessboard to his king and requested that the king give him in exchange one grain of rice for the first square on the board, two grains for the second square, four grains for the third, and so forth.

The king agreed and ordered rice to be brought from his stores. The fourth square on the chessboard required eight grains, the tenth square took 512 grains, the fifteenth required 16,384, and the twenty-first square gave the courtier more than a million grains of rice. By the forty-first square, a trillion rice grains had to be provided. The payment could never have continued to the sixty-fourth square; it would have taken more rice than there was in the whole world![1]

Why is this story important?

Well, another characteristic of exponential growth is the speed with which an exponentially growing quantity can reach a fixed limit. Take poisonous blue-green algae, for instance.

Excess nutrients in water can cause increases in blue-green algal cells. Warm the water a little and the increase becomes more rapid. Reduce the water quantity and the effect multiplies again.

So nutrient-laden water in hot summer drought conditions will produce an ideal environment for the exponential growth of blue-green algae.

This occurred dramatically in 1991 when 1000 kilometres of the Darling River turned toxic overnight. Just like that. No warning. One day it was drinkable water, the next day poisoned. That's how fast things can change when exponential growth is involved.

Incidentally, this particular problem may affect many of our depleted reservoirs during the Twenty Thirst Century. Blue-green algae has the capacity to render the water in our dams useless for human and animal consumption – overnight.

So you can see that the water crisis is not just about how much water there is left in our rivers and dams and storages. It's also about the *quality* of that water. As quantity diminishes, quality deteriorates.

Another point to remember is that **exponential growth cannot continue for very long in a finite space with finite resources**. If you want to learn more about exponential growth and its implications, there's a useful chapter on the subject in *Limits to Growth: The 30 Year Update* (2005).

Exponential growth is easier to observe in populations and economies where it's effects eventually become obvious. In ecosystems, stress from exponential demand is less easy to quantify, but its effect can be disastrous.

What, then, is an ecosystem? Think of a complex, finely balanced mechanism, one in which many intricate individual components cooperate to produce the end result.

Think, for instance, of a clock. Not a modern digital version, but one of the old-fashioned wind-up variety. As a child, did you ever take such a clock apart piece by piece and marvel at the intricacy of each tiny brass cog, wheel and spring? It is an entertaining and fascinating experience. The challenge comes when one attempts to put the whole lot back together before one's parents discover the experiment. If one tiny screw or spring is missing, the clock will not work, or if it does, it will never work properly again.

Ecosystems and the processes of nature can be usefully compared to that clock. Every tiny component plays an essential part in the functioning of the whole. Remove or destroy any component and the clock may stop or become unreliable and unpredictable.

"Ecosystem services" is a mechanical and spiritually arid term for those subtle, complex and often unrecognised natural regenerative processes that support our existence and our way of life. It is only recently that environmental science has acknowledged our dependence on these critical factors, but indigenous people everywhere have always understood this. It's something their ancestors and ours learned from bitter and painful experience during the past 5000 years.

It's a simple proposition, really. When you deplete or exhaust the regenerative capacity of your habitat, there are only two choices left: migrate or perish.

The problem is that in a world that places a greater emphasis on economic and industrial priorities, ecosystem services are at the bottom of the list if they appear at all. Even though ecosystem services have no obvious monetary value their contribution is priceless, invaluable and critical. Scientists who attempt to put an economic value on natural services come up with estimates in the trillions of dollars per year, far exceeding the value of the world's economies combined.

Even the conservative *Australia State of the Environment 2001* recognises their importance in its limited fashion:

These processes ... known as ecosystem services ... include soil formation, nutrient cycling, clean water supply, pollination and waste assimilation. Without these ecosystem services the world's economy would grind to a halt.[2]

Note the writer's obsession with the economy. He's another growth freak who sees the world economy as the prime consideration.

But the world's economy is not the only thing that would grind to a halt. Life would become a precarious struggle for existence; scarcity of fresh water, food and energy could lead to internal and external conflict.

A recently leaked Pentagon report predicts "abrupt climate change could bring the planet to the edge of anarchy as countries develop a nuclear threat to defend and secure dwindling food, water and energy supplies". It concludes by stating: "Disruption and conflict will be endemic features of life ... once again warfare would define human life." (*Australian Financial Review*, 4 May, 2004)

Now, having acknowledged our dependence on ecosystems for our security, wellbeing and, ultimately, for our survival, what is the condition of those vital natural assets?

The Living Planet Index is an indicator of the state of the world's ecosystems, compiled by the World Wide Fund for Nature in 2002. It measures the health of these systems by the abundance – or otherwise – of forest, freshwater and marine species.

The index shows an overall decline of about 37% between 1970 and 2000.

More than one-third of the last of the world's forests, wetlands, freshwater resources, kelp beds, fish – all gone, kaput, used up, destroyed.

That's a lot of ecosystem services that won't be delivering the goods anymore – to a global population that still keeps doubling every 40 years.

In other words we can expect at least one-third less capacity, less available wood, fresh water, fish and wildlife. And diminishing food crops, as depleted soil nutrients are not replenished.

And so it goes.

In the words of the old American Indian leader, Chief Seattle, "Man did not weave the web of life: he is merely a strand in it. Whatever he does to the web, he does to himself."

The *New South Wales State of the Environment 2003* report reiterates warnings of "unsustainable rates of resource depletion", citing "unsustainable use of ground and surface water, energy, soils, native vegetation and fish" – all of these ecosystem services are declining across the state and across the country.

At the same time as the population is increasing.

Multiplying the pressure on stressed ecosystems exponentially means collapse is inevitable. It is likely to be sudden and without further warning.

Our obsession with growth is pushing our luck to its limit and jeopardising the future of our children.

So, to return to the issue of how can we diminish this destructive pattern in our everyday existence – the answer is simple.

We can choose to reduce our consumption.

By reducing our demands we can give ourselves and the environment some desperately needed breathing space – and lower our water consumption as well, because everything we use and consume requires water for its growth, creation or manufacture. Everything from house construction to power generation, from the smallest silicon chip to the biggest watermelon, can be evaluated in terms of the water required for its production. This is called "virtual water".

By calculating the amount of virtual water involved we can get some idea of our level of daily dependence, not merely on water, but on cheap water.

Consider the following: the CSIRO estimates that a dollar's worth of rice at the supermarket entailed the use of 7500 litres of water, a dollar's worth of raw cotton took 1600 litres to produce, and a dollar's worth of sugar cane, 1239 litres. Other estimates suggest that producing a cotton t-shirt takes 760 litres of water, while a pair of pure cotton queen-size sheets entail 6000 litres.

For each litre of petrol, 18 litres of water was used and polluted.

Economising on water use obviously reduces the energy necessary to deliver it, but conserving energy also conserves water. Thermoelectric power plants (coal, oil, natural gas, nuclear or geothermal) use up water through evaporation, as excess heat is removed from condensers. Mining the fuels used to run these plants also consumes water, and hydroelectric power generation results in evaporation of water from storage reservoirs.

All together, the water required to service energy demands is substantial – an estimated 8.3 litres per kilowatt hour of delivered electricity.

So next time you get your electricity bill you can compute how many litres of water it took to produce your power.

The less you consume, the less the environment suffers.

It isn't just a matter of our individual consumption either. When our imports and exports are analysed for their virtual water content – for instance, 1 tonne of wheat = 1000 tonnes of water – we discover that Australia is a net exporter of water.

Each year we import approximately 3.5 million megalitres of virtual water, but we export 7.5 million megalitres in our goods and agricultural produce. Which leaves us exporting 4 million megalitres of fresh clean Aussie water every year – more than enough to supply the domestic requirements of all of our capital cities combined.

The driest continent on Earth and we're exporting water!

Is that an appropriate plan for our long-term future, given the dramatic projections of water shortfalls in the national and international supply?

This is yet another illustration of the complexity and contradictions that arise when we examine policy decisions of the past. Irrigators, pastoral, agricultural and horticultural producers are supplied with water at a cost of .05 of a cent per 1000 litres (or even less) in order to keep our export trade flourishing.

We are exporting our lifeblood.

There are plenty of buyers. As the international crisis deepens, water-scarce countries such as Iran, Saudi Arabia, Egypt, Algeria and Yemen are diverting rural supplies to higher-value activities. Industry and cities can pay more for water than farmers, but these countries then have to import grain to offset the resulting loss of production. The world grain market is another indicator that competition for water is beginning to assert itself.

The sensible solution to provide for the future is self-sufficiency.

When energy and oil rise in price, a stage will be reached where global trade will be adversely affected. Transport and material costs will escalate rapidly and those countries and cities that are self-sufficient are the ones that will be least affected.

Water self-sufficiency at a personal level is also a sensible response. Changing rainfall patterns on the east coast of Australia are beginning to mirror those in Perth. Inland catchments are not being inundated: instead, on the east coast, rain is falling on the cities themselves.

This is obvious in Sydney, and on the NSW Central Coast where I live, where dam levels are around 24% of capacity. However, we have had good, heavy rain on the coastal fringe.

Who knows how long this will continue?

Under these circumstances, water tanks are a wise investment. In the 1950s and 60s tanks were so common it was unusual not to have one, even in cities and towns where reticulated water was available.

Municipal councils and water authorities in many areas decided that water tanks represented some sort of imagined threat to their monopoly control of water supplies. It was claimed that tanks were a health hazard, that they harboured mosquitoes and vermin and so on.

For whatever reason, tanks were banned in much of urban Australia. They were torn down and taken to the tip. No one imagined that one day they would need to be resurrected.

That day has come.

Water tanks have suddenly become such big business in Australia that global corporations are buying up small local tank making companies to expand their market share. Expect prices to rise as demand grows and competition diminishes.

Buy now and avoid the rush!

The advantages of collecting and storing your own water are obvious. It provides a level of personal security, while at the same time reducing the amount of run-off and stormwater pollution in cities. An estimated 5 billion litres of stormwater is piped out of our cities and towns every year, while 55% of our incoming household water is used for gardens or flushing toilets. Domestic rainwater tanks of between 5000 and 15000 litres capacity have the potential to reduce mains consumption by as much as 85%.

That's a big win for you, and the environment.

The other aspect of self-sufficiency, is of course, waste. There are two streams of domestic waste – "greywater" and "black water".

Greywater is a term used to describe wastewater from kitchens, bathrooms and laundries; black water is toilet waste.

Greywater accounts for about 45% of household waste – 82 billion litres nationally each year. Around one billion litres of water is used for gardens and flushing toilets. Reusing the greywater in our homes could cut our individual water consumption in half. This would also halve the amount of energy required to deliver water and to pump it away and treat it.

This would, in turn, reduce our greenhouse gas emissions while leaving more water flowing in our rivers and streams.

The most extreme self-sufficiency scenario is one in which rainwater supplies all of the household requirements, and all wastewater is recycled within the house and garden. Rural Australians have been doing this for years, but it's still a novelty in our major cities like Sydney.

In the 1970s there was considerable enthusiasm among environmentally aware architects and engineers for the concept of the "autonomous house" – one that was completely self-sufficient for its energy and water needs. Even though solar power was still in its infancy, it was adequate for the demonstration projects that attracted considerable public and media attention. The Autonomous House at the University of Sydney was one example of a working model.

Now, at last, there is a revival of interest in the idea.

One prominent exponent of the concept is Sydney lawyer Michael Mobbs, whose book *Sustainable House* has been a best-seller in its field. The Mobbs' terrace house in the inner-city suburb of Chippendale has its own independent water, energy and sewerage systems. The additional cost was around $48,000, but Mobbs maintains that it could be accomplished for as little as $20,000. (see www.sustainablehouse.com.au).

When the NSW Premier Bob Carr officially opened the house in December 1996, he described it as "a vision of the future":

> The owners of the property tell me the house will save 102,000 litres of water a year, cut carbon dioxide emissions by around four tonnes per annum and keeps 60,000 litres of sewage out of the waste system. If even a small percentage of homes in the Sydney basin were to use this technology we'd see a massive cut in pollution.

Ten years later, the Mobbs house remains the exception rather than the norm.

Unsustainable resource depletion, an expanding ecological footprint, collapsing ecosystems, increasing population, exponential growth – these are the challenges that the water crisis brings to our urgent attention.

What disturbs me personally is the ignorance, arrogance and confidence of our elected political leaders. This can be a lethal combination at a time of approaching crisis.

I am reminded of the *Titanic*, that giant ship created and designed by experts who pronounced it unsinkable.

When icebergs appeared during its maiden voyage, the captain and the crew were quietly confident. Icebergs didn't frighten them. The ship was unsinkable.

But they could not see that they were approaching a giant iceberg, much of which was hidden underwater. The first-class passengers drinking at the saloon bar hardly noticed the jolt. When the sirens sounded, they could not believe that the threat was real!

Most of them went down with the unsinkable ship.

There was a point at which the master of the *Titanic* could perhaps have avoided the disaster, but the ship's massive size was a disadvantage because the bigger the ship, the more time it takes to change course. A skilful captain needs to be alert enough to anticipate danger so that he can take evasive action as soon as possible.

And here we all are, partying on the giant unsinkable *Titanicus Australis*, with our confident captain John Howard at the helm, heading straight towards the tip of the massive iceberg of global climate change and resource depletion.

And he still hasn't acknowledged its existence.

Life jackets, anyone?

The Ties that Bind

EARLY ONE DARK MORNING, a group of devout Hindus set out in a long boat to cross the Ganges. Their destination was a Shiva Temple where the rituals began at dawn. Tired, cold and hungry, the pilgrims kept their spirits up by smoking a little hashish, a practice not uncommon among devotees in India.

They rowed hard against the strong current for what seemed like hours but still the farther shore seemed as distant as ever. Then the first light revealed their mistake.

In their stoned confusion, they had forgotten to untie the long rope that tethered the boat to the wharf.

I was told this story many years ago by an Indian teacher. The point of the parable was that no matter how much effort and energy we expend in the spiritual or material world, unless we free ourselves from the constraints of false perception and old attachments, we will never reach our destination.

Our attachment to the destructive and false doctrine of the growth economy is the rope that keeps us immobile on the great moving river of history. In spite of all our efforts, no matter how much we struggle, how hard we work and strive, we are still far from shore.

And now there are more passengers. The boat is becoming overloaded, there's not much water left to go round and turbulent waves are lapping at the gunnels.

A storm is coming.

How prepared are we to weather that storm?

There are still those who remain attached to the idea that climate change and global warming are figments of the scientific imagination.

They should read the newspapers.

Records are being broken out there!

On 2 February 2005, Melbourne had a flood unlike any in its recorded history. In 24 hours 120 millimetres of rain fell on the city. Wild winds flattened trees like matchsticks while rising floodwaters swept two tonne manhole covers and wheely bins into creeks and rivers. The overloaded sewage system added raw sewage to the mix. Damage estimates ran into the millions.

And how much of this raging torrent found its way into Melbourne's dams?

Nine days supply.

The rest went out to sea.

This is becoming a repetitive weather pattern along Australia's east coast; from Townsville to Melbourne, rain is falling on cities but not in their emptying inland catchments.

Another newsflash.

In April 2005, Cairns and its surrounds narrowly escaped being hit by a hurricane, a category 5 cyclone, which could have levelled the city, killing thousands. Cyclones develop over the Pacific in response to changes in currents and temperatures. As the ocean warms, more category 5 cyclones are predicted. A lot of beach houses and seaside apartments around Cairns are now up for sale.

Their owners know what sort of damage a future category 5 cyclone can do!

Further south – more broken records.

In the Murray-Darling Basin, farmers were still praying for rain. Some districts have had the highest temperatures and the lowest rainfall on record. April 2005 was the hottest month ever in much of Australia.

National Climate Centre climatologist Blair Trewin said climate change had forced the centre to adjust its predictions of extreme weather.

> The sort of month you might have previously expected to see once in 2000 years, the chance of that has dropped to one in 20 years," he said.

> "We found an increase in background average temperature by 1C increased the chance of a month as extreme as this April by 100 times."

April set 150 high temperature records nationwide and came at the end of a record-breaking four-month period: January 1 to the end of April was the hottest start to a year since Australia-wide data collection began in 1950. (*The Weekend Australian* 7-8 May 2005)

If rain does not fall in the Murray-Darling Basin soon, basic food crops are likely to fail – and meteorologists are saying there's a 50% chance of another year or two of drought.

I don't want this to sound like a doomsday scenario, to use the words of the NSW Auditor-General, but, unless a radical paradiam shift takes place, it could well become one.

Meanwhile, out there in the real world, it's business as usual.

Forty kilometres north of here, bulldozers are busy levelling kilometres of coastal sand dunes at The Entrance on the Central Coast. Watching all this frantic activity, I thought of the Lewis Carroll poem:

The Walrus and the Carpenter were walking on the strand,

They wept like anything to see such quantities of sand.

"If something could be done with it, it surely would be grand."

Developers are building yet another forest of grand apartment blocks just above the high tide mark, apartments with ocean views to blend in with all of the other high rise apartment blocks in the area.

The Walrus and the Carpenter would be thrilled.

Sand dunes? Who needs 'em?

But the thousands of new residents who will occupy this mini city, most of whom are "sea change" refugees from nearby Sydney's hectic lifestyle, what will they drink?

The local water supply is down to 24% capacity and falling. Nevertheless, Gosford Council continues to confidently promote and encourage growth and development; crowding more people in to share a shrinking water supply in spite of opposition from existing residents who have nothing to gain and everything to lose.

This scenario is being repeated in cities and towns across the country.

Why are politicians and councillors and planners always the last to see the obvious?

And why do we continue to elect them when they perform so badly?

Can we trust these people to navigate our perilous path through the Twenty Thirst Century?

What are the alternatives?

If we really want to cut the rope that binds us, or alter the course of the *Titanicus Australis*, we'd better do something about it soon.

Time is running out … and so is our water.

Survival Technology

THIS SECTION IS DEVOTED to the nuts and bolts of water self-sufficiency for those who wish to explore that option. There are plenty of resources available.

An invaluable companion is Michael Mobbs' bestselling *Sustainable House* (Choice Books, $38.50) web: www.sustainablehouse.com.au. Then there's *The Water-Efficient Garden* (Water-Efficient Garden-scapes, $27) and *Waterwise House and Garden – A Guide to Sustainable Living* (CSIRO Publishing, $29.95).

Tanks

Rainfall patterns are changing.

More rain is now falling on coastal cities and less is inundating the inland catchments. So it makes sense to catch and store as much of this as you possibly can – by making sure that some of those pennies from heaven end up in your own rainbank.

When it comes to rainwater tanks there's lots of handy reference material available for free, starting with the Commonwealth enHealth's *Guidebook on the Use of Rainwater Tanks*, a 66 page comprehensive compendium with everything you need to know about tanks. Download it on www.health.gov.au and type 'rainwater tanks' in the search box.

Water suppliers, councils and government departments, all provide useful information about rainwater tanks, rebates and greywater systems on their websites.

www.actewagl.com.au
www.sydneywater.com.au
www.uprct.nsw.gov.au
www.brisbane.quld.gov.au
www.sawater.com.au
www.dhhs.tas.gov.au
www.melbournewater.com.au
www.watercorporation.com.au(WA)
www.savewater.com.au(Vic).

Then go to www.rainharvesting.com.au and get their free 16 page product catalogue, *Ten Steps to Rain Harvesting Sustainable Water,* which contains lots of useful technical advice on tanks, first flush diverters and other helpful information.

The Yellow Pages is, of course, a mine of sources of conventional tanks and equipment but there are some other new products that can save space by fitting into narrow eave spaces or acting as fences.

The Waterwall is a modular freestanding rainwater storage system made from food grade polyethelene. Each module holds 1200 litres and weighs about 1.3 tonnes when full. Available in a range of five colours, the modules can be interconnected to form a fence, partition, screen or feature wall. A reinforced concrete footing is necessary for stabilisation. There is also a slim version designed to fit nearly under the eaves (www.waterwall.com.au).

RainReviva is a water storage system in the form of flexible plastic sacs which can be stored under the house and connected to existing plumbing systems. The sacs can be rolled or folded to fit through narrow openings or under houses with low crawl spaces. (www.rainreviva.com.au).

Bluescope Steel manufactures a range of oval slimline undereave tanks suitable for drinking water. Made from colourbond steel with a food grade polymer lining (Aquaplate), models range in size from 55-4800 litres. (1800 654 774).

If you have the space available or you are designing a new home, then underground concrete or polyethelene tanks could be a viable option.

Contact Enviro Friendly Products (0431 457 091 or web: www.enviro-friendly.com) or Tankmasta (1800 658 265) for further information and costs.

As you will discover, the cost of tanks and rainwater systems varies widely, depending on size, complexity, material etc. Above ground tanks start at $200 for a 500 litre model up to $2000 for a 9000 litre rain bank. That's without installation of course. Below ground tanks cost between $250-$500 per 1000 litres of storage (not including excavation and overburden disposal).

Choose your plumber carefully. Some plumbers have little experience in installing rainwater tanks and may not be familiar with the potential problems or the appropriate guidelines. This can lead to pressure irregularities, incorrect installation which does not meet regulation requirements and/or overcharging.

Greenplumbers is a national organisation of tradesmen who have undergone additional training in water efficient technology, solar hot water installation and water and energy auditing. (greenplumbers.com.au).

Greywater

If you want to learn more about greywater use, The University of NSW Ecoliving Centre offers information, courses, and working models of greywater recycling (www.ecoliving.cat.org.au), while Grey-H2O offers tailored greywater systems and advice (www.EPC-web.com/waterwise.htm).

Greywater can be used untreated but it's subject to many restrictions. One solution is to filter it through a carbon filter and use an ultraviolet lamp to kill any bugs (see water filters for details).

If you're going to recycle washing machine water on your garden then you'll need to use a detergent that's not going to harm the plants. *Aware* and *Planet Arc* laundry powders contain no phosphate or petrochemicals and are safe for greywater use. They are available from supermarkets (planetark.com/gardensafe).

The Greywater Gardener stores and automatically drip feeds washing machine water to the garden. Available from Waterwise Systems, (1300 133 354, www.waterwisesystems.com).

Aqua Nova is an aerated wastewater recycling system which treats all household and toilet waste using naturally occurring bacteria and enzymes. Basically it's an upmarket septic tank adapted to irrigate gardens. It has NSW Health Department accreditation. (www.everhard.com.au).

And while we're on the subject of gardens you may want to think about doing away with the lawn.

The average irrigated lawn uses about 38,000 litres during the hot summer months; gardeners also lavish 10 times more pesticide per hectare of turf than farmers use on crops. Fertilisers and other chemicals not taken up by plants and grass seep down to contaminate underground water supplies.

Big water savings can be made by planting native drought-resistant grasses, groundcovers, wild-flowers and other plants that thrive naturally in your local climate.

Drip or trickle irrigation can legally be used during all but the most severe water restrictions. Drippers work at low water pressure and use small openings to wet a very specific soil area. Drip irrigation reduces both evaporation and garden water consumption by delivering water only where it's needed (www.everydripcounts.com.au and www.aquasava.com.au).

Water filters

If self sufficiency is not an option for whatever reason, at the very least you need to think seriously about getting a water filter.

Why?

It is an axiom that as water quantity diminishes, quality inevitably deteriorates. Australia's surface water was never in very good shape at the best of times. During prolonged droughts when levels in dams and storages drop below 50% that water quality, already questionable, rapidly declines.

Think of big dams like Sydney's Warragamba as giant puddles. As consumption and evaporation reduce the volume of the puddle, concentrating dirt and contaminants into the water that remains, its clarity and purity are compromised.

At the moment it looks as if many of us will be drinking from shrinking puddles for some time to come.

In order to kill bugs in the remaining water, chlorine dosing is increased. Increased chlorine dosing of contaminated water leads to higher levels of carcinogenic by-products arising from the disinfection process. To eliminate these, and the taste and smell of chlorine you need a filter.

Domestic water purifiers have become essential items of household technology in Australia. This is not simply because of a general lack of confidence in the quality of tap water; there is also a growing recognition that our ever-increasing rate of water consumption, the vast quantities of water required, and rising levels of catchment pollution and pipeline contamination have inevitably led to diminished water quality. Even the best treatment plant in the world can only reduce pollutants to a tolerable level, because of the volume of water involved.

It is safer and certainly more reliable to remove the last traces of contaminants from drinking water at the tap. The decision to invest in water treatment depends on the level of concern you have with the quality of your tap water. This will influence the choice of equipment necessary to purify it.

There are several treatment technologies that can be used individually or in combination. The principal types marketed in Australia use activated carbon, carbon block, reverse osmosis, ceramic filters, distillation, or ultraviolet light. These vary from point-of-use devices, which treat water at one source only, to point-of-entry devices, which are installed so that they treat all water entering the household plumbing.

There are four points to consider:

- the quality of the tap water to be treated
- the contaminants you want to eliminate
- how much you are prepared to spend – prices vary from $30 to $2000, depending on the degree of purification provided
- the frequency and ongoing costs of maintenance.

An important factor in your choice is the range of contaminants you wish to remove or reduce. The words 'remove' and 'reduce' need to be carefully considered in relation to water treatment. In practical terms it is difficult (if not impossible) to remove 100% of any substance or impurity found in water. For this reason most water filters have a specified percentage reduction for specific contaminants.

Now let's briefly examine the different treatment methods.

How filters work

Filters are composed of a substance that traps, adsorbs, or modifies pollutants in the water that flows through them. This substance is called a medium. There are many different types of filter media. Some mechanically trap pollutants with an ultra-fine sieve or strainer, while others use a process called absorption, in which contaminants are retained within the microscopic pores of the medium.

The rating of a water filter or purifier tells you what size particles it will and won't remove. Filters are rated in micro metres or microns. A micron is one millionth of a metre. A human hair is 70 microns in diameter, a Cryptosporidium oocyst 4–6 microns, and a Giardia oocyst 8–12 microns.

There are two types of filter ratings: nominal and absolute. A nominal rating indicates the smallest particle size that the filter should remove or reduce, in accordance with its design criteria. It is an estimated value, not a precise one. A 5 micron nominal filter, for example, should trap 95% of all particles 5 microns or larger. An absolute filter rating, on the other hand, refers to a certified reduction level, usually 99.9%. Therefore a 5 micron absolute filter will remove 99.9% of particles 5 microns or more in diameter.

Sediment filters

Sediment or particulate filters are fine sieves that trap dirt and other particles. Using one as a pre-filter will protect a water purifier from damage and extend its life, because it will take longer to become clogged with muck.

Sediment filters range from coarse to fine, and are rated accordingly. They can be made from wound string, rigid foam or pleated film, and are usually mounted under the sink. The life of a sediment filter depends on the rubbish in the water; 6–12 months is average.

Activated carbon filters

Activated carbon filters are particularly effective at removing pollutants that create unpleasant taste, colour, and odour in water. These fast-acting filters can eliminate or reduce the levels of chlorine byproducts, pesticides, herbicides, and other organic and industrial chemicals.

Activated carbon is made from porous organic material such as coal, coconut, lignite, and wood. When these are activated by exposure to high temperatures in the absence of oxygen, the result is a substance with millions of microscopic pores and a vast surface area; half a kilo of activated carbon provides more than 50 hectares of surface with the capacity to cling to or adsorb smaller organic molecules.

There are two forms of carbon in general use: granular and block. Carbon granules are the size of coarse sand, while carbon block is finely powdered carbon compressed into a solid mass. Carbon or carbon block filters with a rating of 1 micron absolute will remove organic contaminants and a high percentage of pollutants (including protozoan cysts); their life span will be extended if a sediment filter is used.

The effectiveness of activated carbon depends on the type of carbon, the amount used, the filter design, how slowly the water flows through it (contact time), and the water quality. A slow flow-rate is desirable, as this allows more impurities to be adsorbed.

There are several kinds of activated carbon filters. Make sure you assess the rating and capacity of a filter before you buy it – read the specifications, and compare these with other brands.

Some activated carbon filters may provide an environment for the growth of bacteria, which feed on the trapped organic material. If the municipal water-supply disinfection is functioning properly and the filter is cleaned and changed regularly, then only non-pathogenic bacteria are likely to be present. These can be flushed away by allowing the filtered water to flow for 30 seconds before the first use each morning.

To get the most out of a carbon filter, it should be kept free of sediment and heavy organic impurities by the incorporation of a sediment filter as an integral part of the system design. It is imperative that filter cartridges be replaced regularly before they reach their expiry date, rather than after.

Ceramic candle filters

These are effective against bacteria, parasites, and sediments. Some models can filter down to 0.9 of a micron absolute. The candle has a hollow core of unglazed porcelain that can be scrubbed with a soft brush or steel wool when cleaning becomes necessary. Flow rates vary according to the model. Some ceramic filters are fitted with an additional activated carbon core to increase their taste and odour-reduction efficiency. The use of a sediment filter is recommended to extend the life of the ceramic candle and to keep cleaning to a minimum.

Reverse osmosis purifiers

Reverse osmosis (RO) is commonly used to provide pure water for the food, mining and pharmaceutical industries, as well as for domestic consumption.

Osmosis is a process that occurs when two solutions of different concentrations are separated by a semi-permeable membrane. Water passes through the membrane in the direction of the more concentrated solution.

RO water purification works by forcing the water under pressure against an ultra-fine, semi-permeable membrane designed to allow single water molecules to permeate through, while at the same time rejecting most contaminants. The membrane acts as a mechanical filter, straining out particulate matter, micro-organisms, asbestos, and even single molecules of heavier organic compounds.

A typical RO purifier consists of three filters in series, plus a storage tank. The first is a sediment filter to remove particulates, the second housing contains the ultra-fine RO membrane, and the third is an activated-carbon cartridge to remove any remaining chlorine byprod-ucts such as chloroform or THMs.

Once the contaminants are separated from the clean water, they are washed away so that they don't build up in the purifier. Such a system removes a wide spectrum of impurities; the only energy required is that of mains-water pressure. RO removes turbidity, sediment, colloidal matter, total dissolved solids, toxic metals, radioactive elements, pesticides, and herbicides. This can have significant health benefits. Trials at the University of Quebec in Canada in 1993 concluded that 'individuals consuming the un-modified tap water experienced gastro-intestinal symptoms 30% more frequently than those who consumed the water after reverse osmosis treatment'.

The efficiency of RO depends on the pressure and temperature of the water. Most domestic RO units operate with a minimum water pres-sure of 250 kPa. A typical system produces water at a slow rate – almost drop by drop – so most have a pressurised storage tank (11 litres) and a separate dedicated tap installed over the sink. Water drawn from the tap comes from the storage tank. The RO unit then slowly refills the tank. The average system produces about 40 litres per day, more than enough for an average family.

Because an RO unit is self-cleaning, it may take 1–3 litres of water to produce one pure litre. However, the amount of rejected water neces-sary to purify the few litres required for drinking each day is hardly substantial. The average domestic RO unit will use about 40 litres per day to flush away contaminants – average household consumption is around 1000 litres per day.

Unlike filters, RO membranes don't accumulate pollutants, but the membranes themselves gradually degrade with use. While the sediment and carbon filters will probably need replacement every 6–12 months, membranes should be changed every 3–5 years, or as specified by the manufacturers.

Ultraviolet disinfection

When bacteria and other micro-organisms are a problem, UV treatment is an excellent technology. The device contains a radiation lamp giving off a level of radiation that kills bacteria. It is also effective against algal toxins. The lamp is enclosed in a protective quartz or UV transparent sleeve that allows water to flow around it for effective disinfection.

A disadvantage is that it does not work well with cloudy or dirty water, since the UV radiation cannot reach the bacteria. In these cases, a pre-filter is essential for efficient functioning, and the fitting may need frequent cleaning.

Distillation

Distillation is a simple and reliable method of removing pollutants. The water is boiled to produce steam which, when cooled, condenses back into pure water. Any substances that cannot evaporate are left behind in the boiling chamber. Distillation removes a wider variety of pollutants from water than any other single method of purification.

The only contaminants that distillation does not significantly reduce are volatile organic compounds with a higher boiling point than water. In well-designed units these are removed by the incorporation of a carbon filter.

Distillation, like reverse osmosis, is a slow process. Purified water is produced at the rate of around 1–2 litres per hour. The electrically powered units consume power – the estimated cost is between 7–10 cents per litre of water. Unlike water filters there is no decrease in performance over time. A 10-year-old distiller will produce the same quality of water as it did when new. However, they do need regular cleaning and maintenance in order to continue operating efficiently.

Selecting equipment

At first there appears to be a confusing variety of makes and models of water-treatment devices, but these can be divided into five basic groups: jugs, counter-top models, under-sink equipment, whole-house filters and shower filters.

Jugs

In September 1993 *Choice Magazine* tested seven different filter jugs. All used carbon filters, but some had added ion-exchange resin for additional efficiency. However, with the exception of an expensive model, which had the best performance, all were rated equally efficient at removing taste and odour contaminants, suspended solids, copper, and chlorine.

Countertop models

Countertop purifiers or filters are freestanding units connected to the sink tap. If you are considering one, avoid those with rubber push-on connections; these have an irritating habit of blowing off under increased water pressure. The best models have a diverter, which screws onto the end of the tap. When the button on the diverter is pulled, water flows through the purifier. Reverse-osmosis, activated carbon, carbon block, ion-exchange, distillation and ceramic filters are all available in countertop form.

Under-sink units

Under-sink installations are convenient because they don't take up valuable kitchen bench space, and there are no tubes connected to the sink tap set. Everything is out of sight, and the purified water is piped through a dedicated tap mounted on the sink. Under-sink purifiers are usually larger than countertop units, provide more comprehensive treatment, and require competent installation, preferably by a qualified plumber.

Whole-house filters

Whole-house or point-of-entry filters treat all water coming into the home, whether it is used for drinking, washing, bathing, or for flushing the toilet. They are larger than the under-sink variety. There are three options: a single sediment filter, with or without an additional carbon cartridge, or a water softener.

Shower filters

Inhalation and skin absorption of chloroform and chlorine byproducts is greatest in the shower, where these gases are vaporised. Good ventilation and a shower filter are worthwhile precautions. People with sensitive skin and dry hair may notice a marked improvement in these conditions, since chlorine bonds chemically with the protein in skin and hair, causing dryness, itching, and flaking skin.

Most shower filters have a dual action, incorporating a water-saving head to save hot water, while eliminating chlorine and other contaminants.

Maintenance

All water-treatment devices require routine maintenance; the manufacturer's instructions should be followed carefully, otherwise the device's capacity may be impaired and bacteria may multiply in far greater numbers than if the water were left untreated.

Never try to economise by ignoring your filter's use-by date. It's a good idea to mark this on a calendar in the kitchen as a reminder.

Buying a filter

Home water purification is now big business in Australia, so there's a wide range of equipment to choose from. Some companies market their own devices, while others operate as showrooms offering a variety of makes and models.

Quality water filters are also available from major hardware and plumbing suppliers.

Below is a list of selected manufacturers and distributors of water filters and purifiers. Don't be afraid to ask searching questions or to request details of relevant and meaningful tests of equipment, just as you would for any performance-based appliance.

Further information:

Amway of Australia
46 Carrington Road Castle Hill NSW 2154
Tel: 1300 363 133 Fax: (02) 9843 2169

Aqua One Water Filters and Purifiers
680 Wynnum Road Morningside QLD 4170
Tel: (07) 3395 7122 Fax: (07) 3395 7233
Email: aquaone@gil.com.au

Crystal Clear Purification Systems Pty Ltd
81 Torrens Road Brompton SA 5007
Tel: 1800 353 376

Culligan Australia Pty Ltd
35 Tebbutt Street, Leichhardt NSW 2040
Tel: 1300 655 295 Fax: (02) 9560 1944
Email: info@culligan.com.au
Website: www.culligan.com.au

Cuno Pacific Pty Ltd
140 Sunnyholt Road Blacktown NSW 2148
Tel: 1300 367 362 Fax: (02) 9831 1737
Email: info@cuno.com.au
Website: www.cuno.com.au

Grander Water Technologies (Australia) Pty Ltd
Sole Exclusive Distributor for Australia & Pacific Islands
1863 Gold Coast Highway Burleigh Gold Coast QLD 4220
Tel: (07) 55 687 522 Fax: (07) 55 687 533
Free Call: 1800 675 771 Website: www.grander.com.au
Email: grander@grander.com.au

continues over page

Further information:

Raindance Water Purifiers
140 Sunnyholt Road, Blacktown NSW 2148
Tel: 1800 353 687 Fax: (02) 9676 8428
Email: raindance@ozemail.com.au

T.D. Hatrick & Co
PO Box 5302 93 Ingham Road, Townsville QLD 4810
Tel: (07) 4771 2677 Fax: (07) 4721 1616

Testa Water Purifiers
Lot 36 Norfolk Avenue, South Nowra NSW 2541
Tel: (02) 4423 1477 Fax: (02) 4423 2650
Email: sales@testa.com.au Website: www.testa.com.au

The Water People
PO Box 136 22 High Street, Glen Iris VIC 3146
Tel: (03) 9885 0222 Fax: (03) 9885 6541
Email: waterp@waterpeople.com.au

The Water Shop
425 Miller Street, Cammeray NSW 2062
Tel: (02) 9956 5677 Fax: (02) 9956 5895
Email: filters@watershop.com.au

Waterlogic Australia Pty Ltd
4/25 Howleys Road, Notting Hill VIC 3168
Tel: (03) 9540 0044 Fax: (03) 9543 4323
Website: www.waterlogicaustralia.com.au

NOTES

Flying Blind –The Future of Water in Australia

1. Guang-Guo Ying and Rai Kookana, 'Endocrine Disruption: An Australian Perspective', *Water*, September 2002, p.56.

God Sends a Messenger

1. Donella Meadows, Jorgen Randers and Dennis Meadows, *Limits to Growth: The 30 Year Update*, Earthscan, London, 2005, p.3.

Desal – Great White Hope or Great White Elephant

1. *Meeting the Challenges: Securing Sydney's Water Future. The Metropolitan Water Plan 2004*, NSW Government, Sydney, p.12.

2. Arnold B. Barach, *1975 and the Changes to Come*, Harper and Brother, New York, 1962, p.151.

3. American Water Works Association, *Water Desalting Planning Guide for Water Utilities*, American Water Works Association, John Wiley and Sons, New York, 2004, p.xi.

4. *Platts Global Water Report*, Issue 201, 20 August 2004, p.1.

5. Jeffrey Rothfeder, *Every Drop for Sale*, Penguin Putnam, New York, 2001, p.176.

6. URS Australia, *Economic and Technical Assessment of Desalination Technologies in Australia: With Particular Reference to National Action Plan Priority Regions*, Commonwealth Department of Agriculture, Fisheries and Forestry, Canberra, 2002, p.30. Available from: www.affa.gov.au/content/publications.cfm

7. *Meeting the Challenges*, op.cit., p.12-13.

8. American Water Works Association, op.cit., p.8.

9. Ibid, p.9.

10. Douglas Jehl, *Whose Water Is It?* National Geographic, Washington, 2003, p.207.

11. A.J. McMichael, *Planetary Overload*, Cambridge University Press, Cambridge, 1993, p.148.

12. Ann Young, *Environmental Change in Australia Since 1788*, Oxford University Press, Melbourne, 2000, p.148.

13. Donella Meadows, Jorgen Randers and Dennis Meadows, *Limits to Growth: The 30 Year Update*, Earthscan, London, 2005, p.159.

14. Ibid.

From Toilet to Tap

1. NEWater Promotion, quoted in *Winning Minds to Water Reuse: The Road to NEWater*, Water, March 2004, p.100.

2. Sydney Water, Incident Management Plan, 1997, p.2.

3. Murni Po, Juliane D. Kaercher and Blair E. Nancarrow, *Literature Review of Factors Influencing Public Perceptions of Water Reuse*, CSIRO Land and Water, Technical Report 54/03, December 2003, p.16.

4. Ibid, p.17.

5. S. Khan, A. Shafer, P.Sherman, 'Impediments to Municipal Recycling in Australia', *Water*, March, 2004, p.118.

6. Ibid.

7. H.E. Gibson and N. Apostolidis, *Demonstration, The Solution to Successful Community Acceptance of Recycling*, Water Science and Technology, 43(10), pp 259-266.

Silent Springs and Depleted Aquifers

1. Payal Sampat, 'Uncovering Groundwater Pollution', *State of the World 2001*, Earthscan, London, 2001, p.23.

2. Kate Short, *Quick Poison, Slow Poison: Pesticide Risk in the Lucky Country*, K.Short/Envirobooks, St Albans, NSW, 1994, p.70-1.

3. Quoted in Donella Meadows, Jorgen Randers and Dennis Meadows, *Limits to Growth: The 30 Year Update*, Earthscan, London, 2005, p.166.

4. Australia State of the Environment Committee, *Australia State of the Environment 2001*, CSIRO Publishing, Melbourne, 2001, pp.64-5.

5. Sampat, op.cit., p.41.

No Water Dreaming

1. Jeffrey Rothfeder, *Every Drop for Sale*, Penguin Putnam, New York, p.105.

2. *Sydney Water Drinking Quality Management Plan*, 2000, p.12.

3. Quoted in *Australia State of the Environment 2001*, CSIRO Publishing, Melbourne, 2001, p.99. (see www.infrastructurereportcard.org.au).

Liquid Assets

1. Australia State of the Environment Committee, *Australia State of the Environment 2001*, CSIRO Publishing, Melbourne, 2001, p.59-60.

2. Ibid, p.61.

3. Ibid, p.62.

4. Ticky Fullerton, *Watershed*, ABC Books, Sydney, 2001, p.240.

5. Ibid, p.327.

6. Pratt Water, *The Business of Saving Water*, Pratt Water, Bendigo, 2004, p.6.

7. Fullerton, op.cit., p.339.

Breathing Space

1. Donella Meadows, Jorgen Randers and Dennis Meadows, *Limits to Growth: The 30 Year Update*, Earthscan, London, 2005, p.21.

2. Australia State of the Environment, *Australia State of the Environment 2001*, CSIRO Publishing, Melbourne, p3.

BIBLIOGRAPHY AND
RECOMMENDED READING

Archer, John, *Australia's Drinking Water – The Coming Crisis*, Pure Water Press, 2001. Download a copy at www.johnarcher.com.au

Australian Academy of Technological Sciences and Engineering, *Water Recycling in Australia*, Australian Academy of Technological Sciences and Engineering, Melbourne, 2004.

Australian State of The Environment Committee, *Australia State of the Environment 2001*, CSIRO Publishing, Melbourne, 2001.

Barlow, Maude and Clarke, Tony, *Blue Gold: The Battle Against Corporate Theft of the World's Water*, Earthscan, London, 2002.

Clarke, Robin, *Water: The International Crisis*, MIT Press, Cambridge, MA, 1993.

Fullerton, Ticky, *Watershed: Deciding Our Water Future*, ABC Books, Sydney, 2001.

McDonald, Bernadette and Jehl, Douglas (Eds.), *Whose Water Is It? The Unquenchable Thirst of a Water-Hungry World*, National Geographic Society, Washington, D.C., 2003.

McGuire, Bill, *A Guide to the End of the World – Everything You Never Wanted to Know*, Oxford University Press, Oxford, 2002.

McMichael, A.J., *Planetary Overload: Global environmental change and the Health of the Human Species*, Cambridge University Press, Cambridge, 1993.

Meadows, Donella, Randers, Jorgen and Meadows Dennis, *Limits to Growth: The 30 Year Update*, Earthscan, London, 2005.

Mobbs, Michael, *Sustainable House: Living for our Future*, Choice Books, 2001.

New South Wales State of the Environment 2003,
Department of Environment and Conservation, Sydney, 2003.

Ohlsson, Leif (Ed.), *Hydropolitics: Conflicts Over Water as a Development Constraint*, Zed Books, London, 1995.

Pratt Water, *The Business of Saving Water*, Pratt Water, Campbellfield, 2004.

Rothfeder, Jeffrey, *Every Drop for Sale – Our Desperate Battle Over Water in a World About to Run Out*, Penguin Putnam, New York, 2001.

Roddick Anita et al, *Troubled Water: Saints, Sinners, Truths and Lies about the Global Water Crisis*, Anita Roddick Books, Chichester, UK, 2004.

Shiva, Vandana, *Water Wars: Privatization, Pollution and Profit*, South End Press, Cambridge, MA.

Stauffer, Julie, *The Water Crisis: Constructing Solutions to Freshwater Pollution*, Black Rose Books, New York, 1999.

Talking Water: An Australian Guidebook for the 21st Century, Farmhand Foundation, Sydney, 2004.